ELECTRICAL ENGINE
AND TESTING METHODS

ELECTRICAL ENGINEERING PRINCIPLES AND TESTING METHODS

Rhys Lewis

B.Sc.Tech., C.Eng., M.I.E.E.
Senior Lecturer in Electrical Engineering,
Openshaw Technical College, Manchester

APPLIED SCIENCE PUBLISHERS LTD
LONDON

APPLIED SCIENCE PUBLISHERS LTD
RIPPLE ROAD, BARKING, ESSEX, ENGLAND

ISBN: 0 85334 564 3

WITH 116 ILLUSTRATIONS AND 2 TABLES

© APPLIED SCIENCE PUBLISHERS LTD 1973

Printed in Great Britain by Galliard Limited, Great Yarmouth, Norfolk, England.

Preface

This book is written for people undertaking courses leading to technician and technologist status in electrical and electronic engineering. The contents cover fundamentals of ac and dc circuits, including network theorems, three phase ac systems, transformers, dc machines, electronic and other amplifiers and amplifying devices, electrical instruments and principles of testing and testing methods, the latter including the basic essentials of quality control techniques.

There are two points worthy of note. Firstly, the latter part of the book covers the common testing methods syllabus of the CGLI, excluding reliability, and it is believed that this is the first text to attempt a summary of the requirements of this section of the syllabus, which in the past has caused some difficulty in interpretation to students and teaching staff alike. The second point is that following recent discussion in various technical journals as to the correct way of presenting basic transformer theory, much thought has been devoted to this presentation and this book joins the mere handful at present available which give the correct basic theory.

A number of examination questions are included at the end of each chapter and typical examination papers are given at the end of the book. These should prove invaluable to students revising for the formal examinations.

Acknowledgements

Acknowledgements are due to the following examining bodies for permission to include questions from past papers: The Northern Counties Technical Examinations Council, the East Midland Educational Union and the Union of Educational Institutions. It should be noted that the answers given to questions are those of the author and no responsibility for their correctness or otherwise should be attached to the various examining bodies named.

On a more personal note I would like to thank Miss P. Raymond, Mrs E. Perry and Mrs G. Morgan for their invaluable assistance in the typing and preparation of the manuscript and, as always, my wife for her help and encouragement during the period in which the book was written.

RHYS LEWIS
Manchester

Contents

1.1 Circuits and fields; 1.2 Circuit quantities; 1.3 Inductance; 1.4 Field quantities; 1.5 Interconnection of circuit components; 1.6 Capacitance and resistance in series dc circuits; 1.7 Inductance and resistance in series dc circuits.

2.1 Review of basic theory; 2.2 Equivalent values of a sine wave; 2.3 Phasor diagrams in circuit analysis; 2.4 Opposition to current flow in ac circuits; 2.5 Series circuits; 2.6 Series resonant circuits; 2.7 Parallel circuits; 2.8 Parallel resonant circuits; 2.9 Power in ac circuits; 2.10 Power factor; 2.11 The importance of power factor; 2.12 Power factor improvement.

3.1 Introduction; 3.2 Kirchhoff's laws; 3.3 Thévenin's theorem; 3.4 Norton's theorem; 3.5 Superposition theorem; 3.6 Input and output impedances, matching; 3.7 The maximum power transfer theorem.

4.1 Introduction; 4.2 Methods of connection; 4.3 Delta connected systems; 4.4 Star connected systems; 4.5 Power in a balanced three-phase load.

CHAPTER TEN: TESTING METHODS   235

10.1 Introduction: the need for testing; 10.2 The purpose of a
specification; 10.3 Types of tests; 10.4 Testing and inspection:
sampling; 10.5 Testing techniques: results, tabulation and analysis;
10.6 Testing techniques: types of error; 10.7 Estimation of experimental error; 10.8 Testing of instruments; 10.9 Testing of components; 10.10 Testing of small machines; 10.11 Use of logarithmic
units in testing.

CHAPTER ONE

Fundamentals of dc Circuits

1.1 CIRCUITS AND FIELDS

An electrical circuit is an interconnection of active and passive components. Active components are those which provide energy, *e.g.* batteries, generators, etc. Passive components are those in which the energy from the circuit source is converted to other forms, *e.g.* resistors, capacitors, inductors. There are three kinds of circuit:

(*a*) the conductive circuit in which electric charge moves,

(*b*) the electrostatic circuit in which, after an initial or 'transient' period, charge accumulates and an electric flux is set up.

(*c*) the magnetic circuit in which a magnetic flux is set up due initially to moving electric charge.

A field is the region surrounding or within a component. There are two kinds of field:

(i) an electric field which exists within components in both the conductive and electrostatic circuits.

(ii) a magnetic field which exists within and around components in both the conductive and magnetic circuits.

There are a number of useful circuit and field quantities which, on first acquaintance, may appear confusing. Fortunately, there is an analogy between many of these quantities which aids understanding and memory.

1.2 CIRCUIT QUANTITIES

The three types of circuit are shown in Fig. 1.1. When considering circuits it is helpful to assign a 'cause' and 'effect' to each circuit. For example, in a conductive circuit the 'cause' is voltage and the 'effect' is current. In an electrostatic circuit 'cause' and 'effect' are voltage and electric flux, and in a magnetic circuit they are magneto-motive force and magnetic flux. The ratio of 'cause' to 'effect' gives

1

a measure of the circuit opposition to the 'effect' and we have the quantities *resistance* for the dc conductive circuit and *reluctance* for the magnetic circuit. There is no special name for electrostatic circuit opposition. The ratio of 'effect' to 'cause' we can call circuit 'support' since it is a measure of how easy it is to set up the 'effect'. Circuit support is the reciprocal of circuit opposition. The quantities are *conductance* for the dc conductive circuit, *capacitance* for the

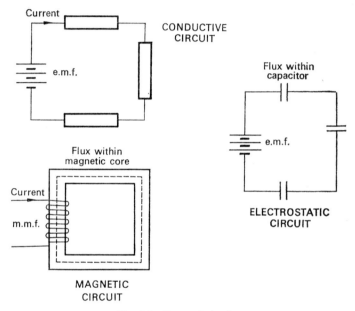

Fig. 1.1   Types of circuit.

electrostatic circuit and *permeance* for the magnetic circuit. All these quantities are shown in Table 1.1, with their units, special names and abbreviations. The reader is advised to make a very careful study of this table in conjunction with the following relevant notes.

Conductive circuit
The cause is loosely given the name voltage. Two more precise names exist—electromotive force (e.m.f.) and potential difference (p.d.). Both quantities measure *energy* per unit *charge* and have the SI unit *joule* per *coulomb*, or *volt*. The term e.m.f. is applied to source voltages and tells us the energy per unit charge which is available from the source. The term p.d. is applied to components or groups of components and, in general, tells us the energy per unit charge

TABLE 1.1 *Circuit quantities*

	Cause	Effect	Circuit opposition cause/effect	Circuit support effect/cause
Conductive circuit	Electromotive Force (e.m.f.) E Volts (V) (1 volt = 1 joule/second)	Current I Amperes (A) (1 ampere = 1 coulomb/ second)	Resistance R ohms (Ω) (1 ohm = 1 volt/ampere)	Conductance G siemens (S) (1 siemens = 1 ampere/volt)
Magnetic circuit	Magneto- motive Force (m.m.f.) F Ampere-turns (A)	Magnetic Flux Φ Webers (Wb)	Reluctance S Ampere-turns/ weber (A/Wb)	Permeance Λ Webers/ ampere-turn (H)
Electrostatic circuit	Potential Difference (p.d.) V Volts (V)	Electric Flux Ψ Coulombs (C)	NO UNIT	Capacitance C Farads (F) (1 farad = 1 coulomb/volt)

which is converted or convertible to other forms of energy within the components. Thus a 6 V e.m.f. means a source which provides 6 joules of energy to each coulomb; a 6 V p.d. across two points in a circuit means that 6 joules per coulomb are being converted or are able to be converted between those points.

The effect is electric current measured in charge units per unit time, *coulombs* per *second*, or *amperes*.

Circuit opposition to current is called *resistance* (for dc circuits) measured in *volts* per *ampere* or *ohms*. Circuit support is called *conductance* (for dc circuits) measured in *amperes* per *volt* or *siemens*. (The siemens was formerly called the *MHO*).

Electrostatic circuit
The cause is as for the conductive circuit. The effect, however, is electric *flux*, which can be regarded as a state of electric stress within the material due to the electric field set up by the voltage.

The SI unit for electric flux is the *coulomb*, which is the same name as the charge unit. One coulomb of flux is that flux associated with one coulomb of charge. Using the coulomb as a flux unit simplifies matters considerably.

Circuit support, the effect/cause ratio, is called *capacitance* and is measured in *coulombs* per *volt*, or *farads*. Capacitance means the ease of setting up of electric flux in an electrostatic circuit. Since electric flux is due to electric charge, capacitance also means the ease of accumulating electric charge within such a circuit. It is this latter meaning which is usually given.

Magnetic circuit

The cause is magnetomotive force (m.m.f.). All magnetism is associated with electric current, including the magnetism of bar magnets which is due to the pattern of spinning electrons within the magnetic material. It is logical then to measure m.m.f. using current. The 'turns' in the unit of m.m.f. means the number of coil turns, since it is found that the magnetic field also depends upon this factor. The size of the turn is unimportant, it is the physical nature of the turn that strengthens the field and thus the flux.

The effect, magnetic flux, is a measure of the stress within the magnetic material due to the field. It must be emphasised that 'lines of flux' both in the electric and magnetic fields are used to illustrate the line of action of the field and have no physical reality. The unit of flux is the *weber*.

The opposition to flux is called *reluctance*, measured in ampere-turns per weber. Since the turn is dimensionless, *i.e.* the size is not important, the unit is abbreviated to amperes per weber. Reluctance is the magnetic circuit equivalent of the conductive circuit quantity resistance. Circuit support measured in webers per ampere is called *permeance*. Conductance, capacitance and permeance are analogous quantities. We shall see later that the quantity inductance is dimensionally the same as permeance and so we can say that conductance, capacitance and inductance are similar quantities.

Notice that the names electromotive force and magnetomotive force are misleading, in that they do not denote forces or 'pressure' as is sometimes wrongly indicated.

The above relationships may be expressed in equation form as follows:

$$V = IR \tag{1.1}$$

$$F = \Phi S \tag{1.2}$$

where V, I and R represent voltage (e.m.f. or p.d.) current and resistance, and F, Φ and S represent m.m.f., magnetic flux and reluctance respectively.

Equation (1.1) is known as Ohm's law.

$$I = GV \qquad (1.3)$$

$$\Phi = \Lambda F \qquad (1.4)$$

$$\Psi = Q = CV \qquad (1.5)$$

where I, V, Φ and F are as above and G, Λ, and C represent conductance, permeance and capacitance respectively. Q represents charge and Ψ electric flux.

The units used in these equations are as indicated in Table 1.1.

1.3 INDUCTANCE

Michael Faraday discovered that, whenever a magnetic field around a conductor is changing, a voltage is induced across the conductor. Lenz discovered that this voltage acts in a direction so as to oppose what is causing it, and so the induced voltage is often called a *back*

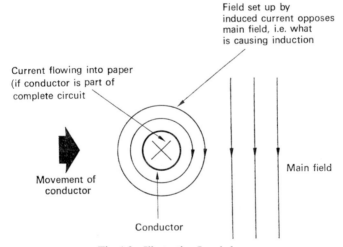

Field set up by induced current opposes main field, i.e. what is causing induction

Current flowing into paper (if conductor is part of complete circuit

Main field

Movement of conductor

Conductor

Fig. 1.2   Illustrating Lenz's law.

e.m.f. This opposition to change shows itself in ways depending upon the circuit set up. One example is illustrated in Fig. 1.2, where the induced voltage causes a current flow which sets up a field acting in the *opposite* direction to the field which originally induced the voltage.

The size of the back e.m.f. depends upon two things, the rate of change of flux with time, and the number of turns of the conductor. We might expect the latter since changing the number of turns changes the strength of the magnetic field. In equation form, Faraday's law is:

$$\text{Back e.m.f.} = -N \times \text{rate of change of flux} \qquad (1.6)$$

where N is the number of turns and the negative sign indicates the opposing nature of the e.m.f.

Since current produces magnetic flux we can relate the back e.m.f. to changing current to give:

$$\text{Back e.m.f.} = -L \times \text{rate of change of current} \qquad (1.7)$$

where L has a value depending on the number of turns and the factors which determine exactly how much flux is obtained per unit of current. The factors affecting the flux produced by electric current will be considered later.

L is called the coefficient of self-inductance, and from eqn. (1.7) its units are *volts* per *ampere* per *second*, or *volt-seconds* per *ampere*. One volt-second/ampere is called one *henry* and is the inductance of a coil which produces a back e.m.f. of 1 volt when the coil current changes at the rate of 1 ampere/second.

By equating the right hand sides of eqns. (1.6) and (1.7), we get

$L \times$ rate of change of current with time
$$= N \times \text{rate of change of flux with time}$$
or
$$L = N \times \text{rate of change of flux with current} \qquad (1.8)$$

Now the rate of change of flux with current has the units *webers* per *ampere*, so the unit of inductance, as well as being written *volt-second* per *ampere*, can also be written *weber-turn* per *ampere*, *i.e.* one henry is one weber-turn/ampere. In Section 1.2 we saw that the unit of permeance is the *weber* per *ampere-turn*, and since 'turns' are dimensionless, *i.e.* their *size* is not important, we can see that permeance and inductance are *dimensionally* the same, *i.e.* their unit is the weber/ampere or henry. Their actual values will differ because the inductance unit has 'turns' in the numerator and the permeance unit has 'turns' in the denominator, but we can see that inductance and permeance depend on the same things, and thus inductance, permeance, capacitance and conductance are analogous quantities, all being a measure of circuit 'support'. The following examples are included to emphasise this similarity between circuit quantities and to consolidate the theory in Sections 1.1 to 1.3.

Example 1.1
(a) Calculate the conductance of a circuit having a p.d. of 10 V and a current of 3 A.
(b) Calculate the current in a resistor of conductance 10 mS when a voltage of 10 J/C is applied.
(c) Calculate the energy dissipated in 5 minutes in a resistor having a p.d. across its ends of 50 J/C and a current of 10 mC/s flowing through it.

Solution

(a) $\qquad$ Conductance $= \dfrac{\text{current}}{\text{voltage}}$ $\qquad$ from eqn. (1.3)

$$= \frac{3}{10} \text{ A/V}$$

$$= 0\cdot3 \text{ S}$$

(b) $\qquad$ Current $=$ voltage $\times$ conductance $\qquad$ from eqn. (1.3)
$$= 10 \times 10 \times 10^{-3}$$
$$= 0\cdot1 \text{ A}$$

(c) Multiplying voltage $\times$ current in SI units gives

$$\frac{\text{joules}}{\text{coulomb}} \times \frac{\text{coulomb}}{\text{second}} = \frac{\text{joules}}{\text{second}} \text{ (or watts)}$$

Multiplying this by time (in seconds) gives the energy dissipated:
$$\text{Energy} = 50 \times 10 \times 10^{-3} \times 300 \text{ J}$$
$$= 150 \text{ J}$$

Example 1.2
In a certain magnetic circuit, a flux of 10 mWb is set up by a current of 1 A flowing through a 500 turn coil. Calculate the permeance of the circuit and thus the coil inductance.

Solution
From eqn. (1.4),

$$\text{permeance} = \frac{\text{flux}}{\text{m.m.f.}}$$

$$= \frac{10 \times 10^{-3}}{500 \times 1} \text{ Wb/A}$$

$$= 20 \text{ }\mu\text{Wb/A}$$

$$= 20 \text{ }\mu\text{H}$$

The unit of inductance is the weber-turn/ampere. Multiplying the permeance unit by $(turn)^2$ yields

$$\left(\frac{weber}{ampere\text{-}turn}\right) \times turns^2 = \frac{weber\text{-}turn}{ampere}$$

i.e. permeance $\times$ turns^2 = inductance.
Thus
$$\begin{aligned}
inductance &= (500)^2 \times 20 \times 10^{-6} \\
&= 5 \text{ weber-turn/ampere} \\
&= 5 \text{ H}
\end{aligned}$$

Example 1.3
Calculate the circuit capacitance if a p.d. of 10 V results in a stored charge of 50 μC.

Since the charge is 50×10^{-6} C the flux is 50×10^{-6} C and from eqn. (1.5),
$$\begin{aligned}
capacitance &= \frac{flux}{voltage} \\
&= \frac{50 \times 10^{-6}}{10} \\
&= 5 \ \mu C/V \\
&= 5 \ \mu F
\end{aligned}$$

1.4   FIELD QUANTITIES

Quantities specifically describing the region within or surrounding circuit components are known as field quantities. They are recognisable by their making reference to the physical size, *i.e.* length, area or volume, of the field region. In the discussion of circuit quantities we assigned a 'cause' and 'effect' for each circuit type and examined the meaning of the ratios of these quantities. This method will be expanded in this section to give a 'cause per unit length' an 'effect per unit area', and we shall go on to examine the meaning of these quantities.

The 'cause per unit length' for a conductive or a static field is the same thing and is called the *voltage gradient* or *electric field strength*. For a magnetic field, the 'cause per unit length' is called *magnetising force* or *magnetic field strength*.

The 'effect per unit area' for the electrostatic field, conductive field and magnetic field is called *electric flux density, current density* and *magnetic flux density,* respectively.

The ratio of 'cause per unit length' to 'effect per unit area' gives a measure of *opposition* between the opposite faces of a unit cube, *i.e.* a cube having sides of unity length, so that the area of all faces is unity and the distance between any two opposite faces is also unity. It has a special name only in the conductive field case and is called *specific resistance* or *resistivity.*

The ratio of 'effect per unit area' to 'cause per unit length' is a measure of the *support* between opposite faces of a unit cube. It is

TABLE 1.2 *Field quantities*

	Cause per unit length	Effect per unit area	Opposition per unit cube $\left(\dfrac{\text{cause per unit length}}{\text{effect per unit area}}\right)$	Support per unit cube $\left(\dfrac{\text{effect per unit area}}{\text{cause per unit length}}\right)$
Conductive field	Voltage gradient E volts/metre (V/m)	Current density J amperes/ metre^2 (A/m^2)	Resistivity ρ ohm-metres (Ω-m)	Conductivity σ siemens/metre (S/m)
Magnetic field	Magnetic field strength H ampere turns/ metre (A/m)	Flux Density B tesla, (T) (1 tesla = 1 Wb/m^2)	NO UNIT	Permeability μ henrys/metre (H/m) (1 henry = Wb/A)
Electrostatic field	Electric field strength E volts/metre (V/m)	Flux density D coulombs/ metre^2 (/Cm^2)	NO UNIT	Permittivity ε farads/metre (F/m)

given the name *conductivity* (specific conductance), *permittivity* (specific capacitance) and *permeability* (specific permeance) for the conductive, electrostatic and magnetic fields respectively. The word 'specific' implies that the quantity is being applied to a part of the field having given dimensions.

All these quantities are summarised in Table 1.2 which gives special names, units and abbreviations. Explanatory notes are given below.

Conductive field

Voltage gradient or electric field strength is obtained by dividing the voltage across a specific part of the field by the distance across which it acts. It is measured in volts per metre.

Current density is obtained by dividing the current by the area through which it flows. It is measured in amperes per square metre.

The ratio voltage gradient to current density is the specific resistance or *resistivity*. The reciprocal of this quantity is called *conductivity*. The reader is advised to satisfy himself as to how the units given are derived.

Electrostatic field

Voltage gradient is as for the conductive field. Electric flux density is obtained by dividing the electric flux by the area through which it acts. The ratio of electric flux density to voltage gradient is the permittivity. The permittivity ε of a material in which a field is present can be expressed as a multiple of the permittivity of a vacuum, ε_0, as follows

$$\varepsilon = \varepsilon_r \varepsilon_0$$

where ε_r is the *relative permittivity* of the material. An insulator to be used in an electrostatic circuit is often called a *dielectric*, and another name for ε_r is *dielectric constant*.

The dielectric constant for a particular material remains substantially constant over a wide range of electric field strength values.

Magnetic field

The magnetic field strength is obtained by dividing the m.m.f. acting over a length of circuit by the length itself. Magnetic flux density is obtained by dividing the flux in a part of the circuit by the area through which the flux acts. The ratio of flux density to magnetic field strength is the permeability of the material. The permeability μ of a material can be expressed as a multiple of the permeability of a vacuum μ_0 as follows

$$\mu = \mu_r \mu_0$$

where μ_r is the *relative permeability* of the material. The relative permeability for a particular material does not remain constant but is dependent upon the magnetic field strength. This is due to the phenomenon of *saturation*, which makes it increasingly difficult to increase the flux density in a material as the magnetic field strength is increased. The relative permeability thus falls. This is shown in Fig. 1.3.

The above relationships may be expressed in equation form as follows:

$$\frac{I}{a} = \sigma \frac{V}{l} \tag{1.9}$$

$$\frac{\Psi}{a} = \varepsilon \frac{V}{l} \tag{1.10}$$

$$\frac{\Phi}{a} = \mu \frac{F}{l} \tag{1.11}$$

where the symbols are as indicated in Table 1.2 and l and a represent length and area respectively. Using the symbols D, E, B and H, eqns. (1.10) and (1.11) are written

$$D = \varepsilon E \tag{1.12}$$

$$B = \mu H \tag{1.13}$$

Graphs of D against E and B against H are shown in Fig. 1.3(a) and 1.3(b). Notice that Fig. 1.3(a) is a set of linear graphs and 1.3(b) shows the phenomenon of magnetic saturation. Equation (1.9) may be rewritten

$$\frac{V}{l} = \rho \frac{I}{a} \tag{1.14}$$

The value of ρ for a particular material varies with temperature but for most materials is independent of voltage gradient. Certain special materials have been developed in which ρ does vary; these are used mainly in surge protection circuits.

Example 1.4
Calculate the resistivity of a piece of material 10 cm long and 5 cm^2 in area if there is a p.d. of 10 V across it and a current of 0·5 A flows.

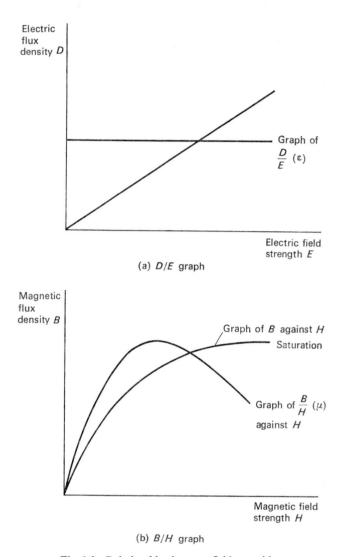

(a) D/E graph

(b) B/H graph

Fig. 1.3   Relationships between field quantities.

$$\text{Voltage gradient} = \frac{10}{10 \times 10^{-2}} \text{ V/m}$$

$$= 100 \text{ V/m}$$

$$\text{Current density} = \frac{0 \cdot 5}{5 \times 10^{-4}} \text{ A/m}^2$$

$$= 1000 \text{ A/m}^2$$

Hence, by eqn. (1.14), the resistivity

$$\rho = \frac{100}{1000}$$

$$= 0 \cdot 1 \ \Omega\text{-m}$$

(The conductivity is $1/\rho$ which has the value 10 S/m).

Example 1.5
Calculate the capacitance of a piece of dielectric having a dielectric constant 8, length 0·01 cm and area 0·1 cm^2 (assume $\varepsilon_0 = 8\cdot85$ pF/m).

From eqn. 1.10

$$\frac{\Psi}{a} = \varepsilon \frac{V}{l}$$

therefore

$$\frac{\Psi}{V} = \varepsilon \frac{a}{l}$$

but $\Psi/V = C$, from eqn. (1.5).
Thus

$$\text{capacitance} = \frac{8\cdot85 \times 10^{-12} \times 8 \times 0\cdot1 \times 10^{-4}}{0\cdot01 \times 10^{-2}}$$

$$= 7\cdot08 \text{ pF}$$

Example 1.6
Calculate the current required in a 500 turn coil to set up a flux of 10 mWb in a piece of material which is 5 cm long, has an area of 1 cm^2 and a relative permeability at this value of flux density of 5000 ($\mu_0 = 0\cdot4\pi \ \mu$H/m).

From eqn. 1.11

$$\frac{\Phi}{a} = \mu \frac{F}{l}$$

Thus

$$F = \frac{\Phi l}{a\mu}$$

$$= \frac{10 \times 10^{-3} \times 5 \times 10^{-2}}{1 \times 10^{-4} \times 5000 \times 4\pi \times 10^{-7}} \text{ ampere-turns (A)}$$

$$= 797 \text{ A}$$

Hence since current I is given by

$$I = \frac{F}{N}$$

where N is the number of turns

$$I = \frac{797}{500} \text{ amperes}$$

$$= 1 \cdot 59 \text{ A}$$

1.5   INTERCONNECTION OF CIRCUIT COMPONENTS

The circuit illustrated in Fig. 1.4(a) is a series connection of resistors R_1, R_2 and R_3, having potential differences V_1, V_2 and V_3 and a common current I.

Total p.d.

$$V_{tot} = V_1 + V_2 + V_3$$
$$= IR_1 + IR_2 + IR_3$$

from eqn. (1.1).

If

$$V_{tot} = IR_{tot}$$

where R_{tot} is the total equivalent resistance, then

$$IR_{tot} = IR_1 + IR_2 + IR_3$$

and

$$R_{tot} = R_1 + R_2 + R_3 \tag{1.15}$$

For the parallel circuit illustrated in Fig. 1.4(b) in which resistors R_1, R_2 and R_3 have individual currents I_1, I_2 and I_3 and a common p.d. V

Total current

$$I_{tot} = I_1 + I_2 + I_3$$

$$= \frac{V}{R_1} + \frac{V}{R_2} + \frac{V}{R_3}$$

from eqn. (1.1).
If

$$I_{tot} = \frac{V}{R_{tot}}$$

where R_{tot} is the total equivalent resistance, then

$$\frac{V}{R_{tot}} = \frac{V}{R_1} + \frac{V}{R_2} + \frac{V}{R_3}$$

and

$$\frac{1}{R_{tot}} = \frac{1}{R_1} + \frac{1}{R_2} + \frac{1}{R_3} \qquad (1.16)$$

(a) Series resistor connection

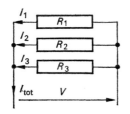

(b) Parallel resistor connection

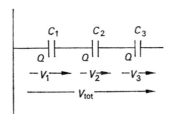

(c) Series capacitor connection

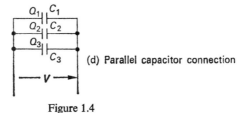

(d) Parallel capacitor connection

Figure 1.4

The circuit illustrated in Fig. 1.4(c) shows a series connection of three capacitors C_1, C_2 and C_3 having voltages, V_1, V_2 and V_3 and a charge on each Q coulombs.

Total voltage

$$V_{tot} = V_1 + V_2 + V_3$$

$$= \frac{Q}{C_1} + \frac{Q}{C_2} + \frac{Q}{C_3}$$

from eqn. (1.5).

If

$$V_{tot} = \frac{Q}{C_{tot}}$$

where C_{tot} is the equivalent total capacitance then

$$\frac{Q}{C_{tot}} = \frac{Q}{C_1} + \frac{Q}{C_2} + \frac{Q}{C_3}$$

and

$$\frac{1}{C_{tot}} = \frac{1}{C_1} + \frac{1}{C_2} + \frac{1}{C_3} \tag{1.17}$$

For the parallel circuit shown in Fig. 1.4(d) the capacitors C_1, C_2 and C_3 are connected in parallel. The p.d. V is common and there is a charge on each capacitor, determined by the p.d. and the capacitance, of Q_1, Q_2 and Q_3.

Total charge

$$Q = Q_1 + Q_2 + Q_3$$
$$= C_1 V + C_2 V + C_3 V$$

from eqn. (1.5) and if $Q = C_{tot} V$, where C_{tot} is the total capacitance, then

$$C_{tot} V = C_1 V + C_2 V + C_3 V$$

and

$$C_{tot} = C_1 + C_2 + C_3 \tag{1.18}$$

Notice that the equations for resistors in parallel and capacitors in series are similar as are the equations for resistors in series and capacitors in parallel. The 'opposite' nature of resistance and capacitance is thus reflected in these equations, in that the equations of 'opposite' states, series and parallel, are similar for 'opposite' quantities.

1.6 CAPACITANCE AND RESISTANCE IN SERIES DC CIRCUITS

The circuit illustrated in Fig. 1.5(*a*) shows a series connection of a capacitor of capacitance C farads and a resistor of resistance R ohms connected via a switch to a dc supply of E volts. The voltage across a capacitor cannot change instantaneously since the capacitor

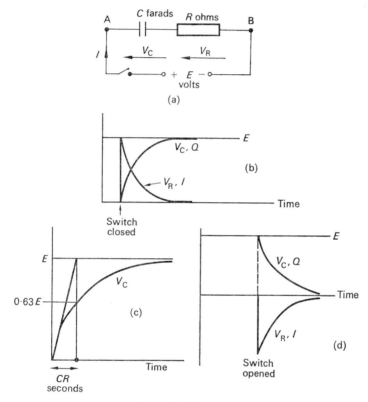

Fig. 1.5 *CR* circuits.

must have time to charge, and charge and voltage are related to one another as was shown in eqn. (1.5). The charge on a fully charged capacitor depends upon the applied voltage and the capacitance. The time taken for this charge to accumulate must depend upon the total charge and thus upon the capacitance of the capacitor. Secondly, how long it takes for the capacitor to acquire the total charge also depends upon the charging current, the greater the current the smaller the time required. It will be recalled that electric current is in fact a measure of charge per unit time. The current in the circuit shown is equal to the voltage across the resistor V_R divided by the resistance R. The current is directly dependent upon V_R since R is fixed.

When the switch is closed there can be no charge on the capacitor at the instant of switching, thus the total voltage E must appear across the resistor R and the charging current at the instant of switching must therefore rise to E/R. This is the maximum possible value of current since once the capacitor begins to charge, its p.d. V_C will rise and V_R, which determines the current, must fall since at all times

$$E = V_R + V_C$$

With the initial surge of charging current the capacitor begins to charge quickly, but as carriers begin to accumulate on the plates the charging process slows down because of the repulsion between carriers and charge which must be overcome by the current.

A graph of charge Q, voltage V_C, voltage V_R and current I plotted against time is shown in Fig. 1.5(b). Notice that the graphs of V_R and I have the same shape and the graphs of Q and V_C have the same shape. Notice also that at all times $E = V_C + V_R$. The rate of charge is reflected in the slope of the charge/time curve and as can be seen it falls off with time.

This type of curve is well known in mathematics. It is called an *exponential* curve and the equation describing it contains the constant e, a constant which, like π, appears again and again in electrical engineering analyses. Tables of the value of e raised to various powers and logarithms of numbers to the base e are usually given in books of standard tables; e itself has the value 2·7183.

The equation of the V_C/time curve is

$$V_C = E[1 - \exp(-t/CR)] \tag{1.19}$$

where t represents time in seconds and where $\exp(-t/CR)$ means $e^{-t/CR}$, i.e. the constant e raised to the power $-t/CR$.

When $t = 0$, $\exp(-t/CR)$ is 1 and $V_C = 0$.

When $t = \infty$, $\exp(-t/CR)$ is 0 and $V_C = E$.

When $t = CR$, $\exp(-t/CR)$ i.e. $\exp(-1)$ is 0·37 and $V_C = 0·63\ E$.

The time $t = CR$ is rather important.

If a tangent is drawn on the V_C/time curve at the point $t = 0$, it cuts the $V_C = E$ line at a time $t = CR$ as shown in Fig. 1.5(c). This shows that CR is the time in which the capacitor would be fully charged if its initial rate of charge remained constant. In fact, it does not and at a time $t = CR$, V_C has reached about 63 per cent of its final value of E volts, as shown.

The time CR seconds is called the *time constant* of the circuit. For CR to be dimensionally equal to seconds, C must be in farads and R in ohms. Theoretically, from eqn. (1.19) V_C never in fact reaches its maximum value of E volts. In practice it is assumed to do so after a time equal to $5 \times$ the time constant, since, at this time,

$$\exp(-t/CR) = \exp(-5)$$
$$= 0 \cdot 001$$

and

$$V_C = 0 \cdot 999 \, E$$

The equations of the other curves are

$$Q = EC[1 - \exp(-t/CR)] \qquad (1.20)$$

$$V_R = E \exp(-t/CR) \qquad (1.21)$$

$$I = \frac{E}{R} \exp(-t/CR) \qquad (1.22)$$

Addition of eqns. (1.19) and (1.21) shows that

$$V_R + V_C = E$$

which is correct. Since $I = V_R/R$ then eqn. (1.22) follows from eqn. (1.21). Equation (1.20) is correct since $Q = CV_C$ and V_C is given by eqn. (1.19).

Notice the form of the equations for an exponentially growing curve such as the V_C or Q curve and for an exponentially decaying curve such as the V_R or I curve.

The maximum value of the variable on the left hand side of the equation is the term immediately before the term containing the exponent e on the right hand side of the equation.

When the switch is opened, the capacitor theoretically will remain charged for ever. In practice it will discharge through the air, its rate of discharge being determined by atmospheric conditions. It is wise when examining circuits containing capacitors to ensure that when the supply is removed, the capacitors are fully discharged before touching. They may be discharged by placing a short circuit across their connecting leads.

For the circuit shown, if the switch is opened and a connection made between points A and B, the capacitor will discharge through the resistor. The exponential curves of the charging process are repeated; only the voltage V_C and the charge Q are now decaying exponentially. The voltage V_R and the current I similarly fall to zero, but since the initial voltage V_R and current I are in the opposite sense to that during charging, the curves are as shown in Fig. 1.5(d) and the variable quantities appear to rise exponentially from a negative value to zero. The equations are:

$$V_C = E \exp(-t/CR) \tag{1.23}$$

$$Q = CE \exp(-t/CR) \tag{1.24}$$

$$V_R = -E \exp(-t/CR) \tag{1.25}$$

$$I = -\frac{E}{R} \exp(-t/CR) \tag{1.26}$$

The time CR is still called the time constant and it is now the time for V_C to fall by about 63 per cent of its original value. This is shown by putting $t = RC$ in eqn. (1.23).
When $t = RC$

$$V_C = E \exp(-1)$$
$$= 0.37\, E$$

The delaying property of capacitors on dc, *i.e.* the way in which an instantaneous rise or fall in voltage can be prevented, is widely used in electronics.

Example 1.7
A direct voltage of 100 V is applied to a circuit containing a 0·1 μF capacitor in series with a 10 MΩ resistor; calculate

(a) the voltage across the capacitor at 0·5 s, 1 s, 3 s and 5 s after closing the circuit,
(b) the initial current in the circuit,
(c) the time constant,
(d) the value of capacitor which must be connected to the 0·1 μF capacitor in order to reduce the charging time to half its present value. How must this capacitor be connected?

Solution
(a) The capacitor voltage at any time t during charging is given by eqn. (1.19) in which $E = 100$ V, $C = 0.1$ μF, $R = 10$ MΩ.

At 0·5 s
$$V_C = 100[1 - \exp(-0.5)]$$
$$= 39\text{ V}$$

At 1 s
$$V_C = 100[1 - \exp(-1)]$$
$$= 63\text{ V}$$

At 3 s
$$V_C = 100[1 - \exp(-3)]$$
$$= 95 \text{ V}$$

At 5 s
$$V_C = 100[1 - \exp(-5)]$$
$$= 99.9 \text{ V}$$

Notice that
$$CR = 0.1 \times 10^{-6} \times 10 \times 10^{6}$$
$$= 1 \text{ second}$$

(b) The initial current is given by E/R and thus equals 10 μA.
(c) Time constant is CR seconds and equals 1 second.
(d) To reduce the charging time by half the capacitance must be reduced by half to 0.05 μF. This can be achieved by connecting a second 0.1 μF capacitor in series with the existing one.

Example 1.8
Calculate the time constant of a circuit made up of the series connection of a capacitor bank consisting of three 100 pF capacitors in parallel and a resistor bank of three 100 MΩ resistors in parallel.

The time constant = capacitance × resistance.
Total capacitance = 300 pF from eqn. (1.18).
Total resistance = 33.3 MΩ from eqn. (1.16).

Then
$$\text{time constant} = 300 \times 10^{-12} \times 33.3 \times 10^{6} \text{ s}$$
$$= 999 \times 10^{-5} \text{ s}$$
$$= 99.9 \text{ } \mu\text{s}$$

Example 1.9
Two capacitors of capacitance 12 μF and 6 μF are connected in series and charged via a 1 MΩ resistor by an applied dc voltage of 100 V. Calculate the voltage across each capacitor after 4 s has elapsed from first closing the circuit.

The total capacitance from eqn. (1.17) is given by
$$\frac{1}{C_{tot}} = \frac{1}{12} + \frac{1}{6}$$

where C_{tot} is in microfarads.
Thus $C_{tot} = 4$ μF, and the
$$\text{time constant} = 4 \times 10^{-6} \times 10^{6}$$
$$= 4 \text{ s}$$

After a time equal to the time constant the voltage across the resistor from eqn. (1.21)

$$= 100 \times \exp(-1)$$
$$= 37 \text{ V}$$

Thus the capacitors share $(100 - 37)$ V between them.

The charge on each capacitor is the same since they are in series, so that if V_1 is the p.d. across the 6 μF capacitor and V_2 is the p.d. across the 12 μF capacitor, the charge Q is given by

$$Q = 6 \times 10^{-6} V_1 = 12 \times 10^{-6} V_2$$

Thus

$$\frac{V_1}{V_2} = 2$$

i.e.

$$V_1 = 2V_2$$

but

$$V_1 + V_2 = 63 \text{ V}$$
$$\therefore 2V_2 + V_2 = 63 \text{ V}$$

and

$$V_2 = 21 \text{ V}$$

thus

$$V_1 = 42 \text{ V}$$

Notice that for each capacitor to have the same charge the smaller capacitor requires the larger voltage. The voltage splits in inverse proportion to the individual capacitances.

1.7   INDUCTANCE AND RESISTANCE IN SERIES DC CIRCUITS

It was stated earlier that self inductance is the characteristic of a circuit of opposing current change. The opposition is due to the fact that, when a current changes, the magnetic field associated with it also changes, and according to Faraday's and Lenz's laws an e.m.f. is induced which tries to prevent the change. On switching on an inductive circuit, the back e.m.f. acts in a direction so as to oppose the rising current. On switching off, the back e.m.f. tries to maintain the falling current. The overall result of this is that there can never be an instantaneous current change in an inductive circuit. It always takes time for the current to settle at the new value, whatever that may be. A graph of induced voltage and current against time is shown in Figs. 1.6(b) and (c) for the circuit shown in Fig. 1.6(a).

It can be seen that the curves have the familiar exponential shape. Notice that the induced e.m.f. has the opposite polarity to the current in all cases. The voltage across the resistor has the same shape as the current since $V_R = IR$ and R is constant [eqn. (1.1)].

As we saw with the curves associated with the CR circuit, there is a time constant associated with exponentially changing voltages

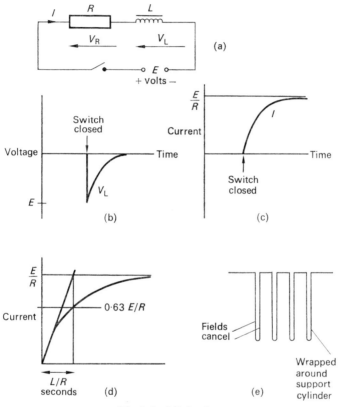

Fig. 1.6 LR circuits.

and currents, being the time in which the levels have changed by approximately 63 per cent of the total. In the series LR circuit the effect of increasing the inductance L increases the time required since L is a measure of the opposition to current *change* within the conductive circuit.

Increasing the resistance, however, reduces the time necessary for the change since it reduces the maximum level of current in the circuit, and thus the current does not have to change by as much as with a smaller value of R. The time constant is, in fact, equal to L/R seconds as shown in Fig. 1.6(d), and the equations describing the relationship of voltage and current for the circuit shown are:

On switching on

$$I = \frac{E}{R} \{1 - \exp\left[-(R/L)t\right]\} \tag{1.27}$$

$$V_R = E\{1 - \exp\left[-(R/L)t\right]\} \tag{1.28}$$

$$V_L = -E \exp\left[-(R/L)t\right] \tag{1.29}$$

On switching off

$$I = \frac{E}{R} \exp\left[-(R/L)t\right] \tag{1.30}$$

$$V_R = E \exp\left[-(R/L)t\right] \tag{1.31}$$

$$V_L = E \exp\left[-(R/L)t\right] \tag{1.32}$$

Notice that the final value of current, which precedes the exponential term in eqns. (1.27) and (1.30), E/R, is determined by the applied voltage divided by the total circuit resistance R. In the theoretical circuit shown no voltage exists across the inductance in the steady-state condition. In practice, of course, the resistance of the inductor is inherent in the coiled wire and it is not possible to separate the resistance from the inductance physically.

The steady-state voltage across an inductive coil is the quantity $V_R = E$ in the circuit and description above. Another point to note is that the back e.m.f. V_L is initially equal to the applied voltage. The back e.m.f. which depends upon the rate of change of current falls as the current rises and its rate of change is reduced.

The inductive effect of a circuit is always present to a greater or lesser degree unless precautions are taken to cancel or neutralise the magnetic field of the conductor. This may be done by winding the conductor back on itself so that the fields cancel (*see* Fig. 1.6(e)), as is done in the manufacture of non-inductive resistors. In circuits associated with magnetic cores, such as motors, generators etc., the inductive effect is enhanced and can cause switching problems, particularly when the circuit is opened. The back e.m.f. may maintain current for a short period via an electric arc across the switch contacts, resulting in damaged contactors. A circuit to prevent this is shown in Fig. 1.7; the back e.m.f. allows a circulating current to flow in the shunt resistor and thus no arc is experienced at the contacts. The switch is arranged mechanically so that the 'off' position is made before the 'on' position is broken.

Use of the inductive arc is made in producing high-voltage sparks for internal-combustion engines, among other applications.

Example 1.10
200 V dc is applied to an inductive coil of inductance 10 H and resistance 500 Ω. Calculate the current in the circuit 10 ms after the circuit is closed. What time elapses before it may be assumed that the steady-state condition exists?

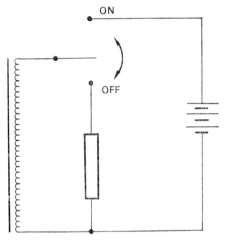

Fig. 1.7   Circuit for reducing arcing. Switch arranged mechanically so that 'off' position is made before 'on' position is broken.

Solution
From eqn. (1.27), the current after 10 ms is

$$= \frac{200}{500}\left[1 - \exp\left(-\frac{500 \times 10 \times 10^{-3}}{10}\right)\right]$$
$$= 0.4 \times [1 - \exp(-0.5)]$$
$$= 0.4 \times (1 - 0.6065)$$
$$= 0.157 \text{ A}$$

Steady-state conditions are assumed to exist after a time equal to 5 × the time constant.

$$\text{Time constant} = \frac{L}{R} = \frac{10}{500}$$
$$= 20 \text{ ms}$$

Steady state may be assumed after 100 ms, *i.e.* 0·1 s.

Example 1.11
The circuit described in Example 1.10 has a switching arrangement whereby a 500 Ω resistor shunts the inductor when the main supply switch is opened. Determine the time for the inductor current to fall to zero when the supply switch is opened.

Solution
The new time constant is determined by the inductance of 10 H and the resistance of 1000 Ω to be

$$\frac{10}{1000} \text{ s}$$

i.e.

$$10 \text{ ms}$$

It will therefore take 50 ms for the current to settle to zero when the supply switch is opened.

PROBLEMS ON CHAPTER ONE

(1) Derive the expression for the total capacitance of a circuit consisting of three capacitors C_1, C_2 and C_3 (*a*) in series, (*b*) in parallel. Three 8 μF capacitors are connected in series, the circuit then being connected across one 4 μF capacitor. Calculate the total capacitance and the total charge accumulated when a direct voltage of 10 V is applied to the complete circuit.

(2) What is meant by *capacitance*? Define the unit of capacitance. A 100 V dc supply is applied to a circuit consisting of an 8 μF capacitor connected in series with a 4 μF capacitor. Determine the voltage across each capacitor when steady state conditions are reached.

(3) What is meant by the power rating of a resistor? Discuss briefly the effect on the total power rating when two resistors of equal rating are connected (*a*) in series, (*b*) in parallel. Determine the maximum voltage that can be safely applied to two 10 kΩ, $\frac{1}{2}$ W resistors connected (*a*) in series, (*b*) in parallel.

(4) Describe with the aid of voltage and current versus time graphs the sequence of events when a direct voltage V volts is applied to a circuit composed of a capacitor C microfarads connected in series with a resistor R ohms. A direct voltage value 100 V is applied to a circuit consisting of a 0·1 μF capacitor connected in series with a 1 MΩ resistor. Calculate: (*a*) the circuit time constant, (*b*) the time

taken for the capacitor voltage to reach 63 V, (c) the time taken for the current to fall to 37 μA, (d) the time taken for steady-state conditions to be reached.

(5) A 50 V dc supply is applied to a circuit consisting of a 4 μF capacitor, a 12 μF capacitor, a 33 kΩ resistor and a 22 kΩ resistor connected in series. Calculate: (a) the time constant, (b) the voltage across the 8 μF capacitor after the voltage is applied.

(6) Explain what is meant by inductance and define the unit. Give one example of an application where inductive effects must be counteracted. An inductive coil of inductance 5 H and resistance 100 Ω is supplied from a voltage source of 20 V dc having an internal resistance of 100 Ω. Calculate the time constant of this circuit and the time taken for the coil current to reach 63 mA.

(7) Using voltage and current versus time graphs illustrate the transient changes when a direct voltage is applied to an inductive coil. A spark supression circuit consists of placing a resistor across a 1 H, 50 Ω coil at the instant of opening the supply switch. Determine the value of this resistor if the decay time of the current is to be reduced to 10 per cent of the value without the resistor. Calculate the new value of decay time.

(8) Define the terms m.m.f. and reluctance when associated with a magnetic circuit. Calculate the reluctance of a steel core of cross-sectional area 5×10^{-4} m^2 if a current of 0·5 A is required in a 500 turn coil to establish a flux density of 0·1 T.

(9) (a) Compare and discuss the quantities conductance, permeance and capacitance. Define the units of these quantities using the International System of Units. (b) Discuss the difference between permeance and inductance and derive a relationship connecting the two quantities.

(10) Explain what is meant by 'relative permeability' as applied to a magnetic circuit. Define the unit of permeability. An iron ring of diameter 0·1 m and cross-sectional area 0·001 m^2 has a coil having 500 turns wound on to it. A current of 1 A through the coil establishes a flux of 0·01 Wb. Calculate the relative permeability of the iron under these conditions. (Absolute permeability $= 0·4\pi$ μH/m).

(11) A 0·1 μF capacitor is charged to 150 V via a resistor of such a value that the charging may be assumed to have a linear relationship with time. If the capacitor is assumed fully charged after 10 seconds, calculate the average charging current.

(12) Describe the initial variations in voltage and current when an inductive coil is supplied with zero frequency voltage. A 10-mH, 50-Ω coil is supplied with 15 V dc. Calculate: (a) the time constant of the circuit, (b) the time taken for the coil current to reach one half the final value, (c) the final value of current.

(13) Define the term 'time constant' when applied to circuits containing (a) inductance and resistance, (b) capacitance and resistance. Determine the value of resistor which, when connected in series with (a) a 1 μF capacitor, (b) a 1 H inductor, will allow the current in each of the two circuits to change by 63 per cent of its maximum value within 50 ms from the instant of applying the supply voltage.

(14) A series of square-wave pulses is applied to a series circuit composed of resistance and capacitance, the output being taken across the capacitor. Compare the output waveform with the input waveform, (a) for a time constant which is short compared with the pulse duration, (b) for a time constant which is long compared with the pulse duration. A series circuit is made up of a 0·1 μF capacitor and a 100 kΩ resistor. Sketch the output pulse appearing across the capacitor on the application to the circuit of a single pulse of width (a) 100 ms, (b) 1 ms.

(15) Explain the term 'relative permittivity' and discuss the effect on the capacitance of a capacitor of changing the dielectric between plates. A parallel-plate capacitor composed of two plates of area 10^{-3} m² situated 10^{-2} m apart has a capacitance of 7 pF. Determine the relative permittivity of the dielectric and the capacitance if this dielectric is then removed. (Absolute permittivity = 8·85 pF/m.)

CHAPTER TWO

Alternating Current Circuits

2.1 REVIEW OF BASIC THEORY

An alternating current or voltage is one which reverses its direction of flow or of action, respectively, at regular intervals. A complete cycle consists of the current or voltage rising to a maximum in one direction then falling through zero to a maximum in the reverse direction, then rising again to zero. The *frequency* of an alternating current or voltage, denoted by f, is the number of complete cycles per second. One cycle per second is called one *hertz*, abbreviated Hz. The time taken for one cycle is called the periodic time and is denoted by T. It follows that

$$T = \frac{1}{f} \tag{2.1}$$

The most commonly used variation in the generation and distribution of alternating currents and voltages for both light and heavy current applications is the sinusoidally varying waveform illustrated in Fig. 2.1. The peak value or amplitude of such a wave is the maximum excursion in either direction; the peak-to-peak value is equal to twice the peak value as shown.

A sinusoidally varying quantity may be derived graphically by plotting a graph of the vertical displacement from the horizontal of

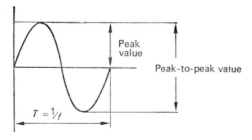

Fig. 2.1 The sine wave.

the end of a rotating arm against either time or angular displacement from the horizontal as the arm rotates. Thus, in Fig. 2.2 the length AP varies sinusoidally as the arm OP rotates. As can be seen from the diagram

$$AP = OP \sin \theta \tag{2.2}$$

and, when $\theta = \pi/2$ radians, $AP = OP$, which is the amplitude or peak value of the generated sine wave. Now if the arm is rotating with an angular velocity of ω rad/s the angle θ is given by

$$\theta = \omega t \tag{2.3}$$

where t is time in seconds, and eqn. (2.2) may be written as

$$AP = OP \sin \omega t \tag{2.4}$$

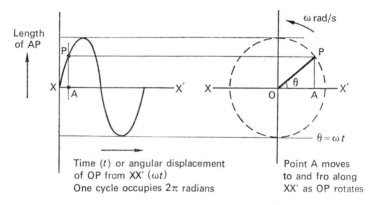

Fig. 2.2   Production of a sine wave by rotating arm.

For each revolution of the arm OP the angular displacement is 2π radians and one complete cycle is plotted. In each second the angular displacement is ω radians and since f cycles are plotted

$$f = \frac{\omega}{2\pi} \tag{2.5}$$

or

$$\omega = 2\pi f$$

so that from eqn. (2.4)

$$AP = OP \sin 2\pi f t \tag{2.6}$$

Thus, the horizontal axis of the sine wave may represent angular displacement ωt or time t as shown in Fig. 2.2.

The length AP represents the instantaneous value of the sinusoidally alternating quantity, the length OP the maximum value so that for a sinusoidally alternating current or voltage, the equations relating instantaneous values to the amplitude or peak values are

$$i = I_m \sin \omega t \qquad (2.7)$$

$$v = V_m \sin \omega t \qquad (2.8)$$

respectively, where i, v represent instantaneous values, I_m, V_m represent peak values, and ω is equal to $2\pi \times$ frequency and is measured in radians per second. The time t or angle ωt is measured from the point at which the sine wave begins its positive excursion, as shown.

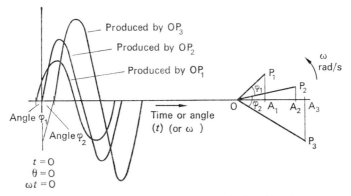

Fig. 2.3   Production of three sine waves displaced in phase.

If three arms OP_1, OP_2, OP_3 situated at fixed angles ϕ_1 between OP_1 and OP_2 and ϕ_2 between OP_2 and OP_3 are rotated at an angular velocity ω, as shown in Fig. 2.3, the vertical displacements from the horizontal of the points P_1, P_2 and P_3, i.e. A_1P_1, A_2P_2 and A_3P_3, will plot three sine waves as shown.

Examination of these sine waves shows that they pass through the same stage or phase at different times or angles. Thus the sine wave produced by OP_2 passes through zero going positive ϕ_1 radians *after* the sine wave produced by OP_1 has passed through this phase, and ϕ_2 radians *before* the sine wave produced by OP_3 passes through this phase. If the time when the OP_2 sine wave passes through zero going positive is taken as $t = 0$, i.e. timing begins at this point, then the equation of this wave is

$$A_2P_2 = OP_2 \sin \omega t \qquad (2.4)$$

as described above. The equations of the OP_1 and OP_3 waves are, respectively,

$$A_1P_1 = OP_1 \sin (\omega t + \phi_1) \qquad (2.9)$$

$$A_3P_3 = OP_3 \sin (\omega t - \phi_2) \qquad (2.10)$$

These equations may be verified by plotting values of A_1P_1, A_2P_2, and A_3P_3 against t or ωt for any fixed value of ω. A brief examination at the point where $t = 0$, *i.e.* where $\omega t = 0$, shows that $A_2P_2 = 0$, $A_1P_1 = OP_1 \sin \phi_1$ and $A_3P_3 = OP_3 \sin(-\phi_2)$ which is verified by the diagram. Also, A_2P_2 is zero when $\sin \omega t = 0$, *i.e.* when $\omega t = 0$, π, 2π, 3π, 4π, etc. radians, A_1P_1 is zero when $(\sin \omega t + \phi_1) = 0$, *i.e.* when $\omega t = -\phi_1$, $\pi - \phi_1$, $2\pi - \phi_1$, $3\pi - \phi_1$, etc. radians, and A_3P_3 is zero when $(\sin \omega t - \phi_2) = 0$, *i.e.* when $\omega t = \phi_2$, $\pi + \phi_2$, $2\pi + \phi_2$, $3\pi + \phi_2$, etc. radians, all of which are shown in Fig. 2.3.

Since the OP_1 sine wave reaches any particular phase before the OP_2 sine wave, it is said to have a *phase lead* of ϕ_1 radians compared to the OP_2 sine wave. Similarly the OP_3 sine wave has a *phase lag* of ϕ_2 radians compared to the OP_2 sine wave. These phase differences may be expressed as angles as above or, less often, in time units by dividing the angles by the angular velocity ω [eqn. (2.3)].

So far, the 'arms' producing these sine waves have been given no special name. They are in fact known as *phasors** since an examination of them when stationary yields the *phase relationship* of the waves they produce. The length of the phasor also yields the amplitude in the diagrams shown so far.

In an ac circuit the various voltages and currents which are distributed about the circuit may have different amplitudes and bear different phase relationships to one another. These may be illustrated by plotting the actual waveforms (voltage or current against time or angle), but since it is known that these waveforms may be produced by the rotation of phasors, as described above, it is more convenient to use these to illustrate magnitude and phase. Extensive use is made of phasor diagrams in ac circuit analysis, and a detailed discussion of them follows in Section 2.3 and the remainder of the chapter.

Example 2.1

A voltage of sinusoidal waveform, amplitude 100 V, leads a current of sinusoidal waveform, amplitude 5 A, by $\pi/6$ radians. Calculate the

---

* Notice that the word 'vector' is incorrect when applied here. The confusion arises because a phasor diagram and a vector diagram both contain lines displaced at angles from one another. However, an angle in a vector diagram implies physical direction, an angle in a phasor diagram implies a time relationship and *not* a physical direction.

instantaneous value of the voltage when the instantaneous value of the current is 2·5 A.

The equation describing the current waveform is

$$i = 5 \sin \omega t$$

and the equation describing the voltage waveform is

$$v = 100 \sin \left(\omega t + \frac{\pi}{6}\right)$$

If the instantaneous value of current is 2·5 A, then

$$2·5 = 5 \sin \omega t$$

and

$$\omega t = \arcsin 0·5$$

$$= \frac{\pi}{6} \text{ rad}$$

so that the instantaneous value of voltage is given by

$$v = 100 \sin \left(\frac{\pi}{6} + \frac{\pi}{6}\right)$$

$$= 86·6 \text{ V}$$

Example 2.2
Calculate the frequency of a sinusoidally varying voltage of amplitude 10 V if its instantaneous value is 7·07 V at a time 2·5 ms after the beginning of a cycle.

The equation of the voltage waveform is

$$v = 10 \sin \omega t$$

and so

$$7·07 = 10 \sin(\omega \times 2·5 \times 10^{-3})$$

Hence

$$\omega \times 2·5 \times 10^{-3} = \arcsin 0·707$$

$$= \frac{\pi}{4}$$

Thus

$$\omega = \frac{\pi}{2·5 \times 4} \times 10^3$$

and since

$$\omega = 2\pi \times \text{frequency}$$

$$\text{frequency} = \frac{\pi}{2\pi \times 2\cdot5 \times 4} \times 10^3$$

$$= 50 \text{ Hz}$$

Example 2.3
Determine the phase angle between two sinusoidally varying currents each of amplitude 1 A if the instantaneous values at a certain particular point in time are 0 and -1 A respectively.

Clearly, when one current is passing through zero the other is at a negative peak and will pass through zero $\pi/2$ rad later. The phase angle is thus $\pi/2$ radians. *See* Fig. 2.4.

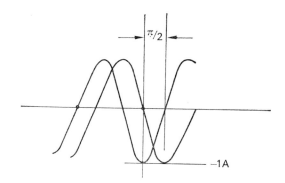

Fig. 2.4   Relating to Example 2.3.

2.2   EQUIVALENT VALUES OF A SINE WAVE

A direct or unidirectional voltage or current has only one value and no problem arises in discussion of such quantities. For a varying quantity the particular value which is being discussed must be specified or a convention must be adopted whereby it is universally understood to which value one is referring. There are five values which are used to describe an alternating quantity. These are:

(i) the peak value or amplitude; as described above this value is the maximum excursion of the voltage or current in either direction measured from zero,

(ii) the peak to peak value; this is the sum of the positive and negative excursions and, clearly, equals twice the peak value,

(iii) the instantaneous value, which is the value at any particular moment in time; this value cannot be used to *define* an alternating quantity since it does not indicate the overall change,

(iv) the average value; as with all averages of time varying quantities, the length of time over which the average is being taken must be stated,

(v) the root-mean-square (r.m.s.) value; this is probably the most commonly used value; it is the value of the equivalent direct current, *i.e.* that value of direct current which if flowing for the same length of time would cause the same energy dissipation.

Values (i), (ii) and (iii) are self explanatory; values (iv) and (v), however, merit further discussion.

Average values

The average value of an alternating quantity which is symmetrical about a fixed axis, *i.e.* either zero or a fixed dc level, is clearly zero if taken over a whole cycle. If taken over half a cycle it then has a value, but it has no practical significance unless the circuit under consideration is being subjected to a succession of half cycles of the same polarity, *e.g.* a rectifier instrument of the type discussed in Chapter 8.

It can be shown by standard mathematical methods, *i.e.* by integral calculus or by graphical techniques, that the average value of a sine wave over half a cycle is related to the peak value by the following equation

$$I_{av} = \frac{2}{\pi} I_m \qquad (2.11)$$

where I_{av} is the average value and I_m the maximum or peak value.

The r.m.s. value

The mean or average value as described above is useful when determining, say, the deflection of an instrument or the effective level once rectification has taken place. However, it should be appreciated that the average value of a sine wave over half a cycle is not the equivalent dc level in an ac circuit. To determine this it is necessary to examine the energy dissipated by an alternating current and evaluate the level of direct current which would dissipate the same energy in the same circumstances (*i.e.* for the same type of load) over the same period of time. The energy being dissipated per second by any current at any time depends not upon the instantaneous level of the current but upon its *square*. Thus, the energy dissipated over a period of time does not depend upon the mean or average of these instantaneous values over this time period but upon the mean of the squares of the instantaneous values. The equivalent direct current is thus the square root of the mean of the squares of the alternating current instantaneous values.

Again, it can be shown mathematically that the r.m.s. value of a sinusoidally alternating current is given by

$$I_{rms} = \frac{I_m}{(2)^{\frac{1}{2}}} \tag{2.12}$$

where I_{rms} is the root mean square value and I_m is the maximum value.

Equations (2.11) and (2.12) apply equally to voltage giving

$$V_{av} = \frac{2}{\pi} V_m \tag{2.13}$$

and

$$V_{rms} = \frac{V_m}{(2)^{\frac{1}{2}}} \tag{2.14}$$

where the subscripts apply as before.

Example 2.4
Calculate the value of direct current which would dissipate the same power in the same load as a sinusoidally alternating current having an instantaneous value given by

$$i = 4\cdot6 \sin 100\pi t$$

From eqn. (2.7), the peak value of this current is 4·6 A. From eqn. (2.12), the equivalent direct current

$$I_{rms} = \frac{4\cdot6}{(2)^{\frac{1}{2}}}$$

$$= 3\cdot25 \text{ A}$$

The *form factor* of any alternating quantity is the ratio of r.m.s. value to average value. For a sinusoidal waveform

$$\text{form factor} = \frac{\pi}{2(2)^{\frac{1}{2}}}$$

$$= 1\cdot11$$

The form factor gives an indication of the wave shape. The more 'peaky' the waveform the higher is the form factor, for a square wave it is unity. The form factor is important in applications where, for example, an indication produced by the average value is used to measure the r.m.s. value. This is discussed in Chapter 8.

2.3 PHASOR DIAGRAMS IN CIRCUIT ANALYSIS

As was stated in Section 2.1 phasor diagrams are used extensively in ac circuit analysis and will be used to clarify the behaviour of all such circuits discussed in this and subsequent chapters. In Section 2.1 it was shown that the vertical displacement of an arm or phasor rotating about a fixed point plotted against either time or the angle between the phasor and the horizontal reference line will yield a sinusoidal wave. Other phasors fixed at one end to the same point will yield other sinusoidal waves displaced from one another by an angle equal to the angle between the appropriate phasors. Thus, provided all the sinusoidally varying voltages or currents in a circuit are at the same frequency, *i.e.* their phasors are rotating at the same angular velocity, the rotating phasors can be stopped, as it were, and their magnitudes and angular displacements from one another will yield the magnitudes and phase displacements of the varying quantities. The distance between the ends of the phasors and the horizontal will be equal to the instantaneous values of the voltages and currents at the instant of 'stopping' the rotation—provided that the phasor lengths are equal to the amplitudes of the sine waves produced. Very often, in practice, the phase relationship and *relative* amplitudes of the various quantities are of more interest than instantaneous values and consequently the phasor lengths are made equal to the r.m.s. values rather than the amplitudes. The lengths must be adjusted if instantaneous values are required.

Other characteristics of phasor diagrams will be dealt with as relevant in the following examination of fundamental circuits.

2.4 OPPOSITION TO CURRENT FLOW IN AC CIRCUITS

The opposition to alternating current flow is measured in the same way and in the same units as opposition to direct current flow, *i.e.*, by dividing voltage by current, the unit being the ohm. However, the situation is somewhat more complex than in the dc case since there are two additional factors—frequency and phase.

AC circuit opposition may be divided firstly into two categories, frequency insensitive and frequency sensitive. Frequency insensitive opposition is the familiar quantity resistance, symbol R. In general, except at very high frequencies where one encounters the 'skin effect', resistance remains the same throughout the useful frequency spectrum. It is determined by dimensions and atomic structure as was discussed in Chapter 1. Resistance also has no effect on phase relationships, voltage and current remaining in phase at all times. A phasor diagram for a purely resistive circuit may be depicted as in Fig. 2.5(a), the current phasor lying on top of the voltage phasor. In this diagram OI_R is the current magnitude (either peak or r.m.s., usually the latter) and OV_R the voltage magnitude.

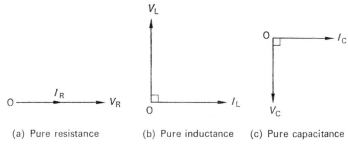

(a) Pure resistance   (b) Pure inductance   (c) Pure capacitance

Fig. 2.5   Phasor diagrams for pure components.

The second form of ac opposition is frequency sensitive and yields a phase shift between voltage and current. This form of opposition is called *reactance*, symbol X, and, as before, is measured in ohms. There are two kinds of reactance, that due to inductance, called inductive reactance, symbol X_L, and that due to capacitance, called capacitive reactance, symbol X_C. The opposition of a circuit containing both resistance and reactance is called *impedance*, symbol Z and this is also measured in ohms. Because of the phase shift introduced by reactive components, impedance is not the algebraic sum of resistance and reactance. This is explained below after inductive and capacitive reactance have been examined more closely.

Inductive reactance is the opposition to alternating current due to inductance. It should be noted that all conductors are inductive to some extent (unless specially wound), but the effect is considerably increased by winding in coil form and including a magnetic core. The device is then known as an inductor. Inductance is, by definition, opposition to current change due to a voltage being induced by the changing magnetic field associated with the current. The greater the rate of change the greater is the induced voltage and thus the opposition to current change. In ac circuits the current is changing continually and the resultant opposition is the inductive reactance. It follows that inductive reactance is proportional to frequency and it can, in fact, be shown that

$$X_L = 2\pi f L \qquad (2.15)$$

where the symbols X_L and f are as previously defined and L is the inductance in *henrys*. The phase change in a pure inductor, *i.e.* a theoretical situation involving a resistanceless conductor, is due to the fact that the applied voltage is at all times equal to the induced voltage, and the induced voltage is proportional to the *rate of change* of the current. If the current has a sinusoidal waveform, the rate of change curve, *i.e.* a curve plotting instantaneous slope against time or angular velocity, has a cosinusoidal waveform, *i.e.* a sinusoidal wave having a phase lead of $\pi/2$ radians over the current waveform. The phasor diagram is shown in Fig. 2.5(b) in which V_L is the voltage applied to the pure inductor and I_L is the current flowing. V_L and I_L are related by the equation,

$$V_L = I_L X_L \qquad (2.16)$$

In practice it is not, of course, possible to have a pure inductor, and for theoretical studies a coil is presumed to contain a pure inductance in series with a pure resistance. This is considered later.

Capacitive reactance is the opposition to alternating current due to capacitance. Again, all conductors exhibit capacitance at all frequencies but the effect is negligible, except at very high frequencies, unless special arrangements involving the provision of 'plates' and a suitable dielectric (to strengthen the electric field for a given voltage) are made. The device is then known as a capacitor. It should be emphasised that current does not actually flow *through* a capacitor but flows only in the conductors connecting it to the supply and to and from the capacitor plates. The current charges or discharges the capacitor in this manner and current flows only when the capacitor is charging or discharging. In a dc circuit a capacitor is eventually either fully charged or fully discharged and circuit current ceases. In an ac circuit, however, the capacitor is constantly undergoing a charging or discharging process and current flows all the time. The greater the rate of change of voltage, *i.e.* the higher the frequency, the larger is the charging/discharging current. Capacitive reactance is thus inversely proportional to frequency and can, in fact, be shown to be given by the equation

$$X_C = \frac{1}{2\pi f C} \qquad (2.17)$$

where the symbols X_C and f are as described above and C is the capacitance in farads. The phase shift due to a pure capacitance, *i.e.* assuming an infinite series resistance, is due to the fact that since charge is proportional to capacitance × voltage (*see* Chapter 1) rate of change of charge is proportional to capacitance × rate of change of voltage, but rate of change of charge (measured in coulombs/second) is current, and so current is proportional to capacitance × rate of change of voltage.

This is the opposite situation to that of the inductor, and so a sinusoidally varying voltage will produce a cosinusoidally varying current, *i.e.* a current varying sinusoidally leading the voltage by a phase angle $\pi/2$ radians. The phasor diagram is shown in Fig. 2.5(*c*) in which I_C, V_C represent capacitor (circuit) current and applied voltage respectively. V_C and I_C are related by the equation

$$V_C = I_C X_C \qquad (2.18)$$

2.5   SERIES CIRCUITS

When a resistive and reactive component are connected in series the resultant phase shift between applied voltage and current is neither zero or $\pi/2$ radians but lies between the two, depending upon the respective effects of each component. For a highly resistive circuit the phase shift is very small, for a highly reactive circuit it is large and almost $\pi/2$ radians. Phasor diagrams for series RL and RC circuits are shown in Fig. 2.6. In these diagrams the symbols refer to quantities as already described, V_S being the supply voltage. As

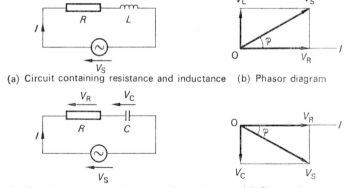

(a) Circuit containing resistance and inductance   (b) Phasor diagram

(c) Circuit containing resistance and capacitance   (d) Phasor diagram

Fig. 2.6   Phasor diagrams for resistive–reactive circuits.

can be seen, V_S is the phasor sum of the resistor voltage V_R and the reactive voltage V_L or V_C. 'Phasor sum' indicates that both magnitude and phase angle are taken into consideration in the summation. The reader may wonder at this point at the reason for the arrows depicting voltage and current in this figure, a practice carried on in succeeding figures of ac circuits. It should be realised that although the same notation is used as for dc circuits (see Chapter 3) the arrows are for identification purposes only, since actual polarities change from instant to instant.

From Fig. 2.6(b)

$$V_S = (V_R{}^2 + V_L{}^2)^{\frac{1}{2}}$$

and since $V_S = IZ$ where Z is the circuit impedance, and from eqns. (1.1) and (2.16), $V_R = IR$ and $V_L = IX_L$ it follows that

$$IZ = [(IR)^2 + (IX_L)^2]^{\frac{1}{2}}$$

so that

$$Z = (R^2 + X_L{}^2)^{\frac{1}{2}} \tag{2.19}$$

Also, if ϕ is the phase angle between supply voltage and supply current,

$$\tan \phi = \frac{V_L}{V_R}$$

$$= \frac{IX_L}{IR}$$

giving

$$\tan \phi = \frac{X_L}{R} \tag{2.20}$$

It also follows that

$$\cos \phi = \frac{V_R}{V_S}$$

i.e.

$$\cos \phi = \frac{R}{Z} \tag{2.21}$$

The significance of $\cos \phi$ is examined later.

Similarly, from Fig. 2.6(d)

$$Z = (R^2 + X_C{}^2)^{\frac{1}{2}} \tag{2.22}$$

$$\tan \phi = \frac{X_C}{R} \tag{2.23}$$

and again,

$$\cos \phi = \frac{R}{Z} \tag{2.21}$$

where Z is given by eqn. 2.22 on this occasion.

The overall effect of a series circuit containing all three types of component is determined by the values of inductance and capacitance and, in particular, the supply frequency. This can be seen by examining Fig. 2.7 which shows graphs of X_L and X_C plotted against a common frequency axis. Since X_L is directly proportional to frequency, this graph is linear and rises as the frequency is increased. The graph of X_C against frequency, on the other hand, is a rectangular hyperbola since X_C is inversely proportional to frequency and falls as the frequency is increased. Now the phase shifting effects of inductive and capacitive reactance are directly opposite to one

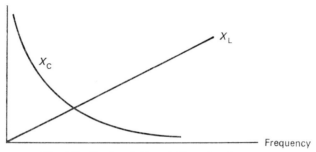

Fig. 2.7   Graphs of X_L and X_C against frequency.

another and the overall effect will depend on the relative effect of each reactance compared to the other. At one frequency the inductive reactance will equal the capacitive reactance and the net phase shift is zero, *i.e.* the circuit appears purely resistive. This phenomenon, termed *series resonance*, has such wide applications it is considered in a separate section following.

2.6   SERIES RESONANT CIRCUITS

Series resonance occurs when the inductive reactance and capacitive reactance are equal, *i.e.* when
$$X_L = X_C$$
and from eqns. (2.15) and (2.17),
$$2\pi fL = \frac{1}{2\pi fC}$$
which yields
$$f^2 = \frac{1}{4\pi^2 LC}$$
and
$$f = \frac{1}{2\pi (LC)^{\frac{1}{2}}} \qquad (2.24)$$

This particular frequency is called the resonant frequency and is usually designated f_r.

At frequencies above resonance the inductive reactive effect is stronger than that due to capacitance, the overall circuit is inductive in nature and the applied voltage leads the circuit current by an amount determined by L, C and R. The phasor diagram is shown in Fig. 2.8(a). The resultant reactive voltage is $V_L - V_C$ in this circuit. From the diagram,

$$V_S = [V_R{}^2 + (V_L - V_C)^2]^{\frac{1}{2}}$$

which yields

$$Z = [R^2 + (X_L - X_C)^2]^{\frac{1}{2}} \qquad (2.25)$$

using $V_S = IZ$ and eqns. (2.16) and (2.18) as before.

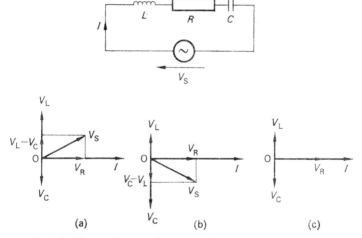

Fig. 2.8   Phasor diagrams for a series circuit containing R, L, C.

At frequencies below resonance the capacitive reactance exceeds the inductive reactance and the circuit is capacitive in nature. The phasor diagram is shown in Fig. 2.8(b). In this case the resultant reactive voltage is $V_C - V_L$ which gives

$$V_S = [V_R{}^2 + (V_C - V_L)^2]^{\frac{1}{2}}$$

yielding

$$Z = [R^2 + (X_C - X_L)^2]^{\frac{1}{2}} \qquad (2.26)$$

At resonance the circuit is resistive and since $X_C = X_L$

$$Z = R$$

The phasor diagram is shown in Fig. 2.8(c), the resultant reactive voltage being zero. Graphs of impedance and current are shown in Fig. 2.9. As can be seen the impedance falls from a high value (capacitive) to a minimum (resistive) and then rises again to a high value (inductive). The current graph has the opposite shape rising to a maximum at resonance.

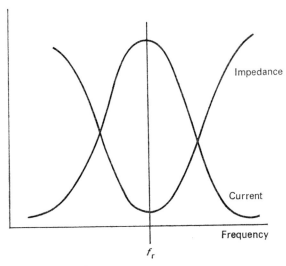

Fig. 2.9   Impedance and current versus frequency for series resonant circuit.

The series resonant circuit is obviously frequency selective in that the current rises to a maximum value at one frequency and falls either side of this frequency. It may be used in any application requiring the ability to respond to a single frequency or small group of frequencies centred about the resonant frequency. Particular applications include tuned amplifiers in radio and television receivers which are required to amplify a narrow band of frequencies centred about the carrier (the electromagnetic wave which carries the audio or video intelligence). An important property of resonant (or tuned) circuits in such applications is the selectivity which is a measure of the response to the resonant frequency relative to frequencies just off resonance.

One measure of selectivity is the Q factor which is defined by the equation

$$Q = \frac{\omega L}{R} \tag{2.27}$$

or since

$$\omega L = \frac{1}{\omega C}$$

at resonance

$$Q = \frac{1}{\omega C R} \tag{2.28}$$

The higher the value of Q the 'sharper' is the current/frequency curve and the better the selectivity. The *bandwidth* of a series tuned circuit is the separation of the frequencies at which the current falls to 0·707 of the maximum value; since power is proportional to the square of the current these points are also those at which the power in the circuit falls to half the value at resonance. It can be shown that Q is also given by the equation

$$Q = \frac{f_r}{B} \tag{2.29}$$

where B is the bandwidth in hertz, and so the higher the value of Q, *i.e.* the greater the selectivity, the larger is the bandwidth for a particular value of resonant frequency. From eqns. (2.27) and (2.28) it can be seen that the selectivity may be increased by increasing the ratio of L/R or by reducing the product of CR. Curves illustrating these effects are shown in Fig. 2.10. If only the resistance of the circuit is increased the maximum value of current is reduced and the selectivity is proportionately reduced. This is also shown in Fig. 2.10.

Example 2.5
A circuit consisting of a 500 Ω non-reactive resistor in series with a fixed capacitor is supplied with 10 mA at a frequency of 1 kHz. The voltage across the capacitor under these conditions is 5 V. Determine the circuit supply voltage and the value of the capacitance.

The voltage across the resistor is $500 \times 10 \times 10^{-3}$ V *i.e.* 5 V. Thus from Fig. 2.6(*d*) which shows the general appearance of the phasor diagram for this type of circuit,

$$V_S = (V_R{}^2 + V_C{}^2)^{\frac{1}{2}}$$

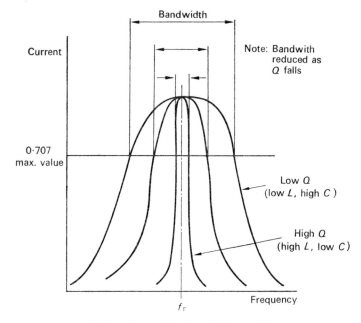

(a) L and C varied; f_r and R constant

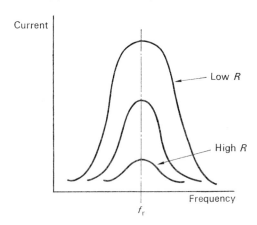

(b) L, C and R constant; R varied

Fig. 2.10   Current/frequency curves as Q is varied.

where V_S is the supply voltage, V_R is the resistor voltage, V_C is the capacitor voltage. Hence, in this case,

$$V_S = (25 + 25)^{\frac{1}{2}}$$
$$= 7{\cdot}07 \text{ V}$$

From eqn. (2.18),

$$X_C = \frac{V_C}{I}$$

$$= \frac{5}{10 \times 10^{-3}} \text{ ohms}$$

$$= 500 \ \Omega$$

From eqn. (2.17),

$$C = \frac{1}{2\pi f X_C}$$

$$= \frac{1}{2\pi \times 10^3 \times 500}$$

$$= 3{\cdot}18 \ \mu\text{F}$$

Example 2.6
If the supply frequency to the circuit of Example 2.5 is reduced to 500 Hz and the applied voltage is correspondingly increased to maintain the same current, determine the new supply voltage.

In Example 2.5,

$$X_C = 500 \ \Omega$$

Since the frequency is halved in this example, X_C is doubled, thus

$$X_C = 1000 \ \Omega$$

From eqn. (2.18),

$$V_C = IX_C$$
$$= 10 \times 10^{-3} \times 1000$$
$$= 10 \text{ V}$$

The resistor voltage remains the same at 5 V and so from the phasor diagram, Fig. 2.6(d),

$$V_S = (5^2 + 10^2)^{\frac{1}{2}}$$
$$= 11{\cdot}18 \text{ V}$$

Example 2.7
The operating coil of a relay having an inductance of 8 mH and a resistance of 30 Ω is connected across an alternating supply voltage

represented by 5 sin 5000 t. Determine the r.m.s. value of the current through the coil and the phase displacement in radians between applied voltage and coil current.

The frequency of the supply is not given directly in this problem but has to be calculated from the voltage expression.

From eqn. (2.8),

$$V = V_m \sin \omega t$$
$$= 5 \sin 5000 \, t \text{ in this case}$$

Hence, the peak value of voltage

$$V_m = 5$$

and

$$\omega = 5000$$

From eqn. (2.5), the frequency

$$f = \frac{\omega}{2\pi}$$

$$= \frac{5000}{2\pi}$$

$$= 796 \text{ Hz}$$

The inductive reactance

$$X_L = 2\pi f L \quad \text{from eqn. (2.15)}$$
$$= 2\pi \times 796 \times 8 \times 10^{-3}$$
$$= 40 \, \Omega$$

Hence, from eqn. (2.19), the impedance

$$Z = (R^2 + X_L{}^2)^{\frac{1}{2}}$$
$$= (30^2 + 40^2)^{\frac{1}{2}}$$
$$= 50 \, \Omega$$

The circuit current is given by dividing voltage by impedance, so that

$$\text{current} = \frac{5 \sin 5000 \, t}{50}$$

$$= 0 \cdot 1 \sin 5000 \, t$$

and from eqn. (2.7),

$$\text{peak current} = 0 \cdot 1 \text{ A}$$

Thus r.m.s. current from eqn. (2.12),

$$I_{rms} = \frac{0.1}{(2)^{\frac{1}{2}}}$$

$$= 0.0707 \text{ A}$$

The phase displacement ϕ from eqn. (2.20) is given by

$$\phi = \arctan X_L/R$$
$$= \arctan 40/30$$
$$= 0.926 \text{ rad}$$

Example 2.8
A series circuit comprising a capacitance of $0.01 \ \mu F$, a coil having an inductance of $2.54 \ \mu H$ and a resistance of $2 \ \Omega$ is fed from a 1 V ac supply. Calculate:

(a) the resonant frequency,
(b) the supply current at resonance,
(c) the voltage across the capacitor at resonance.

(a) From eqn. (2.24),

$$f_r = \frac{1}{2\pi(LC)^{\frac{1}{2}}}$$

thus
$$= \frac{1}{2\pi(2.54 \times 10^{-6} \times 0.01 \times 10^{-6})^{\frac{1}{2}}}$$

$$f_r = 1 \text{ MHz}$$

(b) The current at resonance is equal to the supply voltage divided by the resistance since the reactive components cancel, *i.e.*,

$$\text{current} = \frac{1}{2}$$

$$= 0.5 \text{ A}$$

(c) To determine the capacitor voltage it is first necessary to calculate the capacitive reactance.
From eqn. (2.17),

$$X_C = \frac{1}{2\pi f C}$$

where $f = 1$ MHz, $C = 0.01$ μF. Hence

$$X_C = \frac{10^6}{2\pi \times 10^6 \times 0.01}$$

$$= 15.9 \ \Omega$$

and the capacitor voltage V_C is given by eqn. (2.18):

$$V_C = 15.9 \times 0.5$$

$$= 7.95 \text{ V}$$

An interesting point is made by this example. The reader unfamiliar with ac circuits may wonder how the capacitor voltage can exceed the supply voltage. This is another phenomenon of a series resonant circuit and is termed *voltage magnification*. In the example given there is an equal and opposite voltage across the inductance of 7.95 V and thus the supply voltage appears in its entirety across the resistance. This magnification is possible since it is the *resistance* only of the circuit components which determines the current at resonance. The voltage across each of the reactive components is then determined by the product of the current and the respective reactance. This may or may not exceed the supply voltage depending on current, frequency and the capacitance or inductance. Since the reactive component voltages cancel, none of the magnified voltage appears across the supply terminals. The voltage magnification of a series tuned circuit is used in the input stage of radio receiver circuits as a form of pre-amplification.

From the above example it can be seen that the voltage across the reactive components at resonance

$$V_C = V_L = \frac{I_r}{2\pi f_r C} \qquad \text{(or } I_r 2\pi f_r L)$$

$$= \frac{V_S}{2\pi f_r CR} \qquad \left(\text{or } \frac{2\pi f_r L V_S}{R}\right)$$

where the current at resonance, I_r, is given by

$$I_r = \frac{V_S}{R}$$

V_S being the supply voltage, but from eqns. (2.28) and (2.5),

$$Q = \frac{1}{2\pi f_r CR}$$

(or from eqns. (2.27) and (2.5), $Q = 2\pi f_r L/R$). Hence

$$V_C = V_L = QV_S \qquad (2.30)$$

at resonance. Q is also called the voltage magnification factor for this reason.

2.7 PARALLEL CIRCUITS

As with the series circuit, the overall phase shift between applied voltage and current supplied to a parallel circuit containing resistive and reactive components depends upon the resultant reactance in the circuit. The difference between the circuit types is the usual difference in that for a series circuit the current is common to all components and the voltages may differ, the resultant voltage being

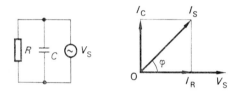

(a) Parallel circuit containing C and R

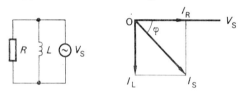

(b) Theoretical parallel circuit containing L and R

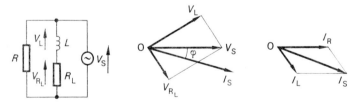

Note: V_{R_L} and I_L are in phase; V_S and I_R are in phase

(c) Practical circuit containing inductor with inherent resistance R_L and resistor

Fig. 2.11 Phasor diagrams for parallel circuits.

the sum of the component voltages, whereas for a parallel circuit the supply voltage is common to all components and the (branch) currents may differ, the resultant current supplied being the sum of the component currents. In ac circuits, of course, 'sum' means phasor sum, *i.e.*, taking both magnitude and phase relationship into consideration.

Phasor diagrams for parallel CR and LR circuits are shown in Fig. 2.11(*a*) and (*b*). It should be noted that the LR circuit consisting of a pure inductance shunting a resistance is not a practical proposition (since an inductor must contain inherent resistance) but it is considered firstly here to aid understanding. In Fig. 2.11(a) I_R, I_C and I_S are the currents in the resistor, the capacitor and from the supply, respectively. I_S is the phasor sum of I_R and I_C. I_R is in phase with the supply voltage V_S, used as reference, I_C leads V_S and thus I_R by $\pi/2$ rad. The resultant phase shift is ϕ between I_S and V_S and, as can be seen, I_S leads V_S.

The phase shift ϕ is given by

$$\tan \phi = \frac{I_C}{I_R}$$

$$= \frac{V}{X_C} \frac{R}{V}$$

i.e.

$$\tan \phi = \frac{R}{X_C} \tag{2.31}$$

Note that this expression is the reciprocal of that obtained for the series circuit, eqn. (2.23).

The impedance of a parallel CR circuit is obtained as follows:

$$\text{capacitor current } I_C = \frac{V_S}{X_C}$$

$$\text{resistor current } I_R = \frac{V_S}{R}$$

$$\text{supply current } I_S = (I_R{}^2 + I_C{}^2)^{\frac{1}{2}}$$

from the phasor diagram of Fig. 2.11(*a*), so that

$$I_S = \left[\left(\frac{V_S}{R} \right)^2 + \left(\frac{V_S}{X_C} \right)^2 \right]^{\frac{1}{2}}$$

$$= V_S \left(\frac{1}{R^2} + \frac{1}{X_C{}^2} \right)^{\frac{1}{2}}$$

and the circuit impedance

$$\frac{V_S}{I_S} = \frac{1}{[(1/R^2) + (1/X_C^2)]^{\frac{1}{2}}} \tag{2.32}$$

In Fig. 2.11(b) the symbols are as before with the addition of I_L, which is the inductor current. Neglecting inductor resistance, I_L lags V_S and thus I_R by $\pi/2$ rad. The phase shift is ϕ with current I_S lagging the supply voltage V_S.

The phase shift ϕ is given by

$$\tan \phi = \frac{I_L}{I_R}$$

$$= \frac{V}{X_L}\frac{R}{V}$$

i.e.

$$\tan \phi = \frac{R}{X_L} \tag{2.33}$$

which is the reciprocal of the expression obtained for a series LR circuit, eqn. (2.20).

The impedance, found similarly to that of a CR circuit, is given by

$$\frac{V_S}{I_S} = \frac{1}{[(1/R^2) + (1/X_L^2)]^{\frac{1}{2}}} \tag{2.34}$$

provided that the coil resistance R_L is neglected. If it is not then X_L in eqn. (2.34) is replaced by Z_L where $Z_L = (R_L^2 + X_L^2)^{\frac{1}{2}}$ and represents the inductor impedance.

If the resistance of the inductor is not ignored the situation becomes a little more complex, yielding the phasor diagram of Fig. 2.11(c). Here V_S is the phasor sum of the voltage across the inductance, V_L, and the voltage across the inductor resistance, V_{RL}. The current in the inductive branch, I_L, is in phase with V_{RL} and is thus not in quadrature with V_S as in the previous circuit. I_S is again the phasor sum of I_R and I_L but the phase shift between I_S and V_S is reduced because of the coil resistance effect in moving I_L away from the quadrature position.

The overall effect of a circuit containing a capacitor in parallel with an inductor depends upon the capacitance, inductance and resistance of the components and particularly the frequency of the supply.

As might be foreseen from the behaviour of the series circuit, there is one frequency at which the reactances are equal, and at this frequency the circuit is resistive. As before, this particular condition is called resonance, and because of the considerable use made of the effect a detailed discussion follows.

2.8   PARALLEL RESONANT CIRCUITS

To facilitate understanding of parallel resonance, first consider a resistanceless coil shunted by a capacitor. The phasor diagrams are shown in Fig. 2.12(a) and (b). In Fig. 2.12(a), the capacitor current I_C (given by V_S/X_C) is greater than the inductor current I_L (given by V_S/X_L). The supply current I_S is the phasor sum of I_C and I_L, namely $I_C - I_L$, and the circuit overall is capacitive. Since X_C is smaller than X_L it follows that the frequency is above resonance. This contrasts with the series circuit which is inductive above resonance. In Fig. 2.12(b) the frequency is below resonance, X_L is smaller than X_C and I_L is larger than I_C. This yields an inductive circuit. At resonance there is no resultant current since I_C equals I_L and thus the supply current is zero. At resonance therefore this circuit presents an infinite impedance. This is only true, of course, because the coil resistance has been neglected. The effects of the coil resistance are discussed below. The circuit current is finite and is a circulating current equal to I_C (and also I_L since $I_C = I_L$). So that although no current is supplied to the ideal circuit, circuit current flows. *Current magnification* takes place.

The phasor diagrams of the practical parallel LC circuit are shown in Fig. 2.12(c), (d) and (e). In this circuit the coil resistance is not zero and a voltage V_{RL} exists across it. In the inductor branch there are two voltages V_L and V_{RL} yielding a phasor sum V_S (*see* Fig. 2.12(c)). The inductor current I_L is in phase with V_{RL} and lags the supply voltage V_S by an angle θ. The capacitor current I_C leads V_S by $\pi/2$ rad and the supply current is the phasor sum of I_C and I_L. The phase shift of the circuit as a whole, *i.e.*, the phase angle ϕ between supply voltage V_S and supply current I_S is determined by the respective magnitudes and phase of I_C, which is determined by V_S/X_C, and I_L, which is determined by V_S/Z_L where $Z_L = (R_L{}^2 + X_L{}^2)^{\frac{1}{2}}$.

Figures 2.12(c), (d) and (e) show the phasor diagrams at frequencies above, below and at resonance. Above resonance I_S leads V_S, *i.e.* the circuit is capacitive, below resonance I_S lags V_S, *i.e.* the circuit is inductive, and at resonance I_S is in phase with the supply voltage. From the phasor diagram, Fig. 2.12(e), it follows that, at resonance,

$$I_C = I_L \sin \theta \qquad (2.35)$$

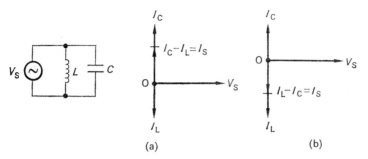

(a)   (b)

Theoretical circuit, inductor with zero resistance

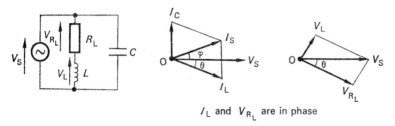

I_L and V_{R_L} are in phase

(c) Practical circuit above resonance

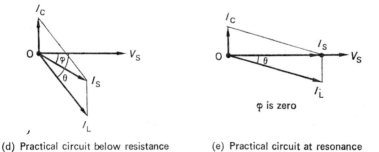

φ is zero

(d) Practical circuit below resistance   (e) Practical circuit at resonance

Fig. 2.12   Phasor diagrams for parallel LCR circuit.

where θ is the phase angle between supply voltage and inductor current given by

$$\tan \theta = \frac{X_L}{R_L} \qquad (2.36)$$

The resonant frequency f_r is determined as follows:

from eqn. (2.35),
$$I_C = I_L \sin \theta$$

from eqn. (2.36),
$$\tan \theta = \frac{X_L}{R_L}$$

and
$$\sin \theta = \frac{X_L}{Z_L}$$

where $Z_L = (R_L{}^2 + X_L{}^2)^{\frac{1}{2}}$, but
$$I_C = \frac{V_S}{X_C}$$

and
$$I_L = \frac{V_S}{Z_L}$$

where
$$Z_L = (R_L{}^2 + X_L{}^2)^{\frac{1}{2}}$$

as before, so that
$$\frac{V_S}{X_C} = \frac{V_S X_L}{R_L{}^2 + X_L{}^2}$$

by substitution in eqn. (2.35). Therefore
$$X_L X_C = R_L{}^2 + X_L{}^2$$

and
$$\frac{L}{C} - R_L{}^2 = X_L{}^2 \qquad (2.37)$$

since
$$X_L = 2\pi f_r L \quad \text{and} \quad X_C = \frac{1}{2\pi f_r C}$$

so that
$$(2\pi f_r L)^2 = \frac{L}{C} - R_L{}^2$$

and
$$f_r = \frac{1}{2\pi}\left(\frac{1}{LC} - \frac{R_L{}^2}{L^2}\right)^{\frac{1}{2}} \qquad (2.38)$$

Unlike the series circuit the resistance of the coil, R_L, affects the resonant frequency of a parallel circuit. If R_L is neglected however, f_r has the same expression as eqn. (2.24):

$$f_r = \frac{1}{2\pi(LC)^{\frac{1}{2}}} \qquad (2.24)$$

In the ideal circuit considered above it was shown that, at resonance, the circuit impedance is infinite. In practice, the coil resistance not being zero leads to a small supply current (Fig. 2.12(e)) at resonance and the impedance is finite but very high.

From Fig. 2.12(e), at resonance, the supply current

$$I_S = I_L \cos \theta$$

$$= I_L \frac{R_L}{Z_L}$$

and since

$$I_L = \frac{V_S}{Z_L}$$

$$I_S = \frac{V_S R_L}{Z_L^2}$$

so that the circuit impedance

$$\frac{V_S}{I_S} = \frac{Z_L^2}{R_L}$$

$$= \frac{X_L^2 + R_L^2}{R_L}$$

and from eqn. (2.37):

$$X_L^2 = \frac{L}{C} - R_L^2$$

yielding

$$\frac{V_S}{I_S} = \frac{L}{CR_L}$$

This value of the circuit impedance is called the *dynamic impedance* of the tuned circuit, symbolised Z_r.

$$Z_r = \frac{L}{CR_L} \qquad (2.39)$$

Impedance/frequency curves of a parallel circuit are shown in Fig. 2.13.

It was shown above that current magnification takes place at resonance, *i.e.* the circuit (circulating) current is greater than the supply current.

The circulating current

$$I_L = \frac{V_S}{Z_L}$$

$$= \frac{I_S Z_r}{Z_L}$$

at resonance

$$= \frac{L}{CR_L} \frac{1}{Z_L} I_S$$

from eqn. (2.39).

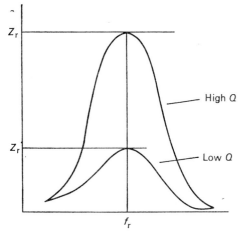

Fig. 2.13 Effect of Q on impedance/frequency curve of parallel resonant circuit.

If R_L is neglected,

$$Z_L \simeq X_L$$

and

$$I_L \simeq \frac{L}{CR_L} \frac{1}{X_L} I_S$$

and since

$$X_L = \omega_r L$$

where

$$\omega_r = 2\pi f_r$$

$$I_L \simeq \frac{1}{\omega_r CR_L} I_S$$

but from eqn. (2.28),

$$Q = \frac{1}{\omega_r C R_L}$$

and thus

$$I_L \simeq Q I_S \qquad (2.40)$$

i.e. the supply current is magnified Q times where Q is as defined earlier. The effect of changing the value of Q on a series tuned circuit was described in Section 2.6. The effect of changing the value of Q in a parallel tuned circuit is quite similar, *i.e.* the higher the value the sharper is the impedance/frequency response curve. The bandwidth is now the separation between the points at which the impedance is 0·707 of the maximum or dynamic impedance but the relationships containing Q still apply. These are repeated below:

$$Q = \frac{\omega L}{R_L} \qquad (2.27)$$

$$Q = \frac{1}{\omega C R_L} \qquad (2.28)$$

$$Q = \frac{f_r}{B} \qquad (2.29)$$

The impedance/frequency curves for various Q values are shown in Fig. 2.13. The low Q curve may be achieved by increasing the coil resistance which reduces the value of Q given by eqns. (2.27) and (2.28) and also reduces the dynamic impedance Z_r given by eqn. (2.37). Altering the coil resistance in this way will also affect the resonant frequency (eqn. (2.36)), but the effect is usually slight.

A shallow response may also be achieved by shunting the circuit as a whole with a resistor. The process, called 'damping', is frequently used in radio receiver circuits when the coil bandwidth is too narrow.

Example 2.9
A 100 mH inductor of resistance 10 Ω is to have a Q factor of 100. Determine the frequencies at which the impedance of a parallel tuned circuit containing this coil drops to 0·707 of the maximum impedance.

The first step is to determine the resonant frequency:

from eqn. (2.27):

$$Q = \frac{\omega L}{R_L}$$

and

$$\omega = 2\pi f_r = \frac{QR_L}{L}$$

hence

$$f_r = \frac{100 \times 10}{100 \times 10^{-3} \times 2\pi}$$

$$= 159 \text{ kHz}$$

From eqn. (2.29), the bandwidth

$$B = \frac{f_r}{Q}$$

$$= 1590 \text{ Hz}$$

The upper end of the bandwidth, $f_r + B/2$, is

$$159\,000 + \frac{1590}{2}$$

i.e. 159 795 Hz.
   The lower end of the bandwidth, $f_r - B/2$, is

$$159\,000 - \frac{1590}{2}$$

i.e. 158 205 Hz.

Example 2.10
A parallel resonant circuit contains a 10 mH inductor having a resistance of 50 Ω. The resonant frequency is 100 kHz. Calculate:

(*a*) the Q factor of the coil,
(*b*) the capacitance of the shunt capacitor,
(*c*) the dynamic impedance, and
(*d*) the capacitor current at resonance, when the supply voltage is 10 V.

(*a*) From eqns. (2.5) and (2.27),

$$Q = \frac{2\pi f_r L}{R_L}$$

$$= \frac{2\pi \times 10^5 \times 10 \times 10^{-3}}{50}$$

$$= 125 \cdot 8$$

(b) From eqn. (2.38),

$$f_r = \frac{1}{2\pi}\left(\frac{1}{LC} - \frac{R_L^2}{L^2}\right)^{\frac{1}{2}}$$

so that

$$\frac{1}{LC} = 4\pi^2 f_r^2 + \frac{R_L^2}{L^2}$$

Clearly, when the values are inserted, the second term, which is of the order of 25×10^6, is negligible compared with the first term which is of the order of $400\,000 \times 10^6$. Approximating, then,

$$\frac{1}{LC} = 4\pi^2 f_r^2$$

and

$$C = \frac{1}{4\pi^2 f_r^2 L}$$

Thus,

$$C = \frac{1}{4\pi^2 \times 10^{10} \times 10 \times 10^{-3}}$$

$$= 267 \text{ pF}$$

(c) From eqn. (2.39), the dynamic impedance

$$= \frac{L}{CR_L}$$

$$= \frac{10 \times 10^{-3}}{267 \times 10^{-12} \times 50}$$

$$= 748 \text{ k}\Omega$$

(d) From eqn. (2.40), the capacitor current (approximately equal to the coil current)

$$I_C = QI_S$$

and

$$I_S = \frac{V_S}{Z_r}$$

$$I_S = \frac{10 \text{ mA}}{748}$$

and

$$I_C = \frac{125 \cdot 8 \times 10}{748}$$

$$= 1 \cdot 68 \text{ mA}$$

Example 2.11

A circuit consisting of a 1 H, 200 Ω inductor in parallel with an 8 μF capacitor is supplied with 100 V at a frequency of 10 Hz below resonance. Calculate the current in each branch of the circuit and hence determine the supply current and the overall phase shift introduced by the circuit.

The phasor diagram of a parallel circuit below resonance is shown in Fig. 2.12(d).

The resonant frequency of the circuit is given by eqn. (2.38):

$$f_r = \frac{1}{2\pi} \left(\frac{1}{LC} - \frac{R_L^2}{L^2} \right)^{\frac{1}{2}}$$

$$= \frac{1}{2\pi} \left(\frac{10^6}{8} - 4 \times 10^4 \right)^{\frac{1}{2}}$$

$$= 46.4 \text{ Hz}$$

The supply frequency is then $46.4 - 10$, *i.e.* 36.4 Hz. The inductive reactance at this frequency, given by eqn. (2.15),

$$X_L = 2\pi \times 36.4 \times 1$$
$$= 228.5 \ \Omega$$

and

$$R_L = 200 \ \Omega$$

Hence, the coil impedance,

$$Z_L = (228.5^2 + 200^2)^{\frac{1}{2}}$$
$$= 304 \ \Omega$$

Hence, the inductor current

$$= \frac{100}{304} \text{ A}$$

$$= 0.329 \text{ A}$$

The capacitive reactance given by eqn. (2.17),

$$X_C = \frac{10^6}{2\pi \times 36.4 \times 8} \Omega$$

$$= 547 \ \Omega$$

Hence, the capacitor current

$$= \frac{100}{547} \text{ A} = 0.183 \text{ A}$$

From the phasor diagram it can be seen that the supply current I_S is the phasor sum of I_L and I_C, but the relationship is not straightforward since I_L is displaced from the reference line by the angle θ. This angle is the phase angle of the coil. The phasor diagram is repeated for clarity in Fig. 2.14.

The phasor I_S may be considered to have two components OX and OY, and

$$I_S^2 = OX^2 + OY^2$$

from Fig. 2.14,

$$OX = I_L \cos \theta$$
$$OY = OZ - YZ$$
$$= I_L \sin \theta - I_C$$

since

$$OZ = I_L \sin \theta \quad \text{and} \quad YZ = I_C$$

Thus

$$I_S^2 = (I_L \cos \theta)^2 + (I_L \sin \theta - I_C)^2$$

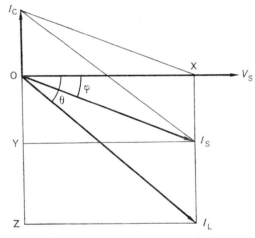

Fig. 2.14   Relating to Example 2.12.

It is now necessary to determine the coil phase angle θ. From eqn. (2.20,)

$$\tan \theta = \frac{X_L}{R}$$

$$= \frac{228 \cdot 5}{200}$$

$$= 1 \cdot 142$$

From tables,

$$\sin \theta = 0.7441$$
$$\cos \theta = 0.6680$$

so that

$$I_S^2 = (0.329 \times 0.668)^2 + (0.329 \times 0.7441 - 0.183)^2$$
$$= 0.0483 + 0.003\ 79$$

thus

$$I_S = 0.228 \text{ A}$$

From the phasor diagram of Fig. 2.14,

$$OX = I_S \cos \phi$$

and

$$OX = I_L \cos \theta$$

Hence

$$I_S \cos \phi = I_L \cos \theta$$

and

$$\phi = \arccos \left(\frac{I_L}{I_S} \cos \theta \right)$$

$$= \arccos \frac{0.329 \times 0.668}{0.228}$$

$$= 0.2653 \text{ rad}$$

2.9   POWER IN AC CIRCUITS

It was shown in Chapter 1 that power in an electrical circuit is determined by the product of voltage and current. In a direct current circuit the power absorbed is constant and the power at any instant of time may be computed directly by multiplying together the voltage and current as read on appropriate instruments. This is because the instantaneous value of direct voltage and current is the same as that read on a voltmeter and ammeter. In ac circuits, however, since the current and voltage fluctuate, the meters employed do not read instantaneous values but are calibrated to read r.m.s. values as already described. Separate voltmeters and ammeters do not of course take into consideration phase shift between voltage and current, and thus the product of voltage and current as read does not necessarily represent true power.

This is clarified in Fig. 2.15 which shows voltage, current and power curves (plotted against time) for various values of phase angle. The dc case shown in Fig. 2.15(a) indicates a constant voltage, current and power value as described above. The ac graphs, shown in parts (b) and (c) for an in-phase and quadrature-phase relationship respectively, are rather more involved, as might be expected. Figure 2.15(b) shows the voltage and current variation with time for a pure resistance. The resultant power variation (determined by multiplying together instantaneous values of current and voltage) lies above the zero axis and, clearly, has a finite positive mean value. This implies a continuously varying dissipation of energy per unit time. It follows that, since the r.m.s. values of alternating voltage and current represent the equivalent values of direct voltage and current in terms of

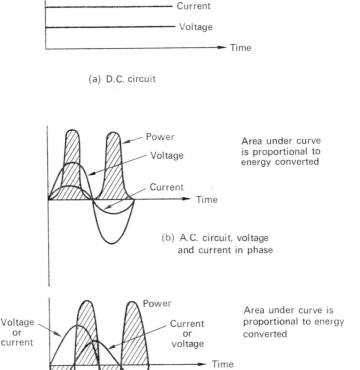

(a) D.C. circuit

(b) A.C. circuit, voltage and current in phase

Area under curve is proportional to energy converted

(c) A.C. circuit, voltage and current in quadrature

Area under curve is proportional to energy converted

Fig. 2.15 Voltage, current and power curves.

the same power dissipation, the mean power dissipated in a pure resistance is given by

$$P = I_{rms}V_{rms} \qquad (2.41)$$

where P is power in watts.

Figure 2.15(c) shows the voltage and current variation for a component which introduces a $\pi/2$ rad phase shift (for an inductance, voltage leading current, for a capacitance, current leading voltage). The power curve, computed as before, is equally displaced about the zero axis and has a zero mean value. The positive half cycles indicate power being taken from the supply, the negative half cycles indicate power being returned to the supply. The net result is zero power dissipation. What is, in fact, occurring is that in a pure inductor the rise of the magnetic field absorbs power which is then returned as the magnetic field collapses. In a capacitor the rise of the electric field absorbs power which is then returned as the electric field collapses. For an inductor the positive power half cycle corresponds to the period when current is rising (since current produces magnetic flux) and for a capacitor to the period when voltage is rising (since voltage produces electric flux). Clearly, in both cases of circuits having a pure reactance eqn. (2.41) does *not* apply, and any attempt to compute power directly from voltmeter and ammeter readings is futile.

Most practical circuits are partly resistive and partly reactive. In such circuits some power is dissipated—in the resistive component—but again eqn. (2.41) applied to the *total* supply voltage and current does not apply. The r.m.s. voltage × current product is in fact multiplied by a factor, called the *power factor*, which has a value lying between zero for a purely reactive circuit and unity for a purely resistive circuit. Power factor is of considerable importance and is discussed in detail below.

2.10   POWER FACTOR

From Section 2.9 it is clear that power is dissipated when voltage and current are in phase but not when these quantities are in quadrature. The immediate question arising is what happens when voltage and current are not in phase, but at the same time are not in quadrature? To answer this it is necessary to re-examine Figs. 2.6(b) and (d), which depict phasor diagrams for resistive–reactive circuits yielding a phase angle ϕ, which lies between zero and $\pi/2$ rad.

It has been shown on a number of occasions that phasors may be summed to yield a resultant phasor using the methods of vector geometry. It follows that any phasor may be considered to be made up of component phasors. This technique was in fact introduced in Example 2.11 to determine the supply current. Thus in Figs. 2.6(b) and (d) which are redrawn in Figs. 2.16(a) and (b) using the r.m.s. values, the current phasor I may be considered to be composed of two components OX and OY. This academic device is in order since

$$I^2 = OX^2 + OY^2$$

and since

$$OX = I \cos \phi$$
$$OY = I \sin \phi$$
$$I^2 = I^2 \cos^2 \phi + I^2 \sin^2 \phi$$
$$= I^2 (\cos^2 \phi + \sin^2 \phi)$$
$$= I^2$$

It should be emphasised that only the one current flows; the splitting into components is a theoretical exercise for convenience.

In both the circuits whose phasor diagrams are depicted by Fig. 2.16, power is dissipated only by in-phase voltage and current, *i.e.* having a value of $V \times OX$. The product $V \times OY$ does not indicate power dissipation.

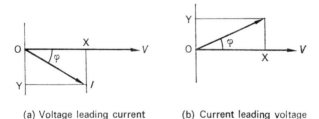

(a) Voltage leading current   (b) Current leading voltage

Fig. 2.16 Phasor components.

The product VI of such circuits is called the *apparent power*, symbol S, and is measured in *volt amperes* (abbreviated VA). The product $V \times OX$, *i.e.* $VI \cos \phi$, is called the *active power*, symbol P, and is measured in *watts* (abbreviated W). The product $V \times OY$, *i.e.* $VI \sin \phi$, is called the *reactive power*, symbol Q, and is measured in *reactive volt amperes* (abbreviated var).

Clearly, since power $= VI \cos \phi$ where ϕ is the phase angle between voltage and current waveforms, $\cos \phi$ is the *power* factor, referred to in Section 2.9. For any circuit having an applied voltage of r.m.s. value V, a supply current of r.m.s. value I and a phase angle ϕ:

$$\text{apparent power } S = VI \text{ in volt amperes} \qquad (2.42)$$

$$\text{active power } P = VI \cos \phi \text{ in watts} \qquad (2.43)$$

$$\text{reactive power } Q = VI \sin \phi \text{ in reactive volt amperes} \qquad (2.44)$$

It follows that

$$\text{power factor } \cos \phi = \frac{\text{active power}}{\text{apparent power}} \qquad (2.45)$$

For a series LR, CR or LCR circuit

$$\cos \phi = \frac{R}{Z}$$

where R ohms is the resistance of the circuit, Z ohms is the impedance of the circuit, given by the appropriate eqns. (2.19), (2.22), (2.25) or (2.26).

Inductive circuits are said to have a *lagging* power factor, capacitive circuits are said to have a *leading* power factor. In both instances the power factor lies between zero and unity, being zero for a purely reactive circuit and unity for a purely resistive circuit.

2.11   THE IMPORTANCE OF POWER FACTOR

Electricity supply authorities charge consumers for the energy which the consumers have used. This energy measured by special meters is determined by the product of power and time and is measured in kilowatt hours (kWh). Clearly, then, the authorities' revenue is determined by the power supplied. This power is determined in turn by the power factor of the load. Whatever the power factor may be, the generating authority must install machines capable of delivering a particular voltage and current even though perhaps not all of the voltage × current product is being put to good use. The generators must be able to withstand the rated voltage and current regardless of the power delivered. For example, if an alternator is rated to deliver 1000 A at a voltage of 11 kV, the machine coils must be wound on a conductor of a cross-section capable of carrying the rated current and having insulation capable of withstanding the rated voltage without breakdown. The apparent power of such a machine is 11 × 1000 kVA, *i.e.* 11 MVA. If the load power factor is unity this 11 MVA will be delivered and used as 11 MW active power and the alternator is being used to the best of its ability. On the other hand, if the load power factor is, say, 0·7 lagging (a common figure for a factory load involving electrical machine drives), only 7·7 MW are taken and provide revenue, but the generator must still be capable of providing 11 kV at 1000 A. The lower the power factor, the worse the situation becomes from the supply authorities viewpoint. Accordingly consumers are encouraged to do all they can to improve the power factor of their load, low power factors often being penalised financially by the supply authorities.

Apart from the financial point of view it is also in the consumers' interest to improve power factor since they too must provide supply cables, from the point of connection to the supply to the actual loading point capable of carrying the load current, whether or not it is being used to full advantage.

2.12 POWER FACTOR IMPROVEMENT

As was stated above the power factor of an individual load is usually lagging due to the inductive nature of the electrical machines which constitute the bulk of the load. It has been demonstrated that capacitive reactance has the opposite effect on phase shift to that of inductive reactance. Consequently, industrial power factor improvement apparatus usually consists of capacitors or capacitive devices, such as, for example, synchronous motors run under certain conditions, connected across the supply so that the capacitive current balances or, at any rate, substantially reduces, the inductive component of current causing the lagging power factor. The determination of the value of capacitance required for particular cases is illustrated in the following examples.

Example 2.12
A 1 H, 200 Ω inductor is connected across a 250 V, 50 Hz supply. Calculate

    (a) the supply current,
    (b) the active and reactive power,
    (c) the power factor,
    (d) the value of the capacitor which when connected across the inductor improves the power factor to unity.

(a) From eqn. (2.15), the inductive reactance

$$X_L = 2\pi \times 50 \times 1$$
$$= 314 \cdot 2 \ \Omega$$

From eqn. (2.19), the inductor impedance

$$Z = (200^2 + 314 \cdot 2^2)^{\frac{1}{2}}$$
$$= 360 \ \Omega$$

The supply current

$$= \frac{250}{360} \ \text{A}$$

$$= 0 \cdot 695 \ \text{A}$$

(b) To determine the active and reactive power it is necessary to find the phase angle.
From eqn. (2.20)

$$\phi = \text{arc tan} \frac{X_L}{R}$$

$$= \text{arc tan} \frac{314 \cdot 2}{200}$$

$$= 1 \cdot 0039 \ \text{rad}$$

From tables
$$\sin \phi = 0 \cdot 8436$$
$$\cos \phi = 0 \cdot 5371$$

From eqn. (2.43), the active power

$$P = VI \cos \phi$$
$$= 250 \times 0 \cdot 695 \times 0 \cdot 5371$$
$$= 93 \cdot 5 \ \text{W}$$

From eqn. (2.44), the reactive power

$$Q = VI \sin \phi$$
$$= 250 \times 0 \cdot 695 \times 0 \cdot 8436$$
$$= 147 \ \text{var}$$

(c) From above the power factor

$$\cos \phi = 0 \cdot 5371 \ \text{lagging}$$

(d) Using the phasor diagram in Fig. 2.16, the reactive current component due to the inductance (OY)

$$= I \sin \phi$$
$$= 0 \cdot 695 \times 0 \cdot 8436$$
$$= 0 \cdot 586 \ \text{A}$$

A capacitor connected across the supply will take a current in antiphase to this. If the capacitor current is equal to the inductor current reactive component, the overall reactive current is zero leaving an in-phase component (OX) equal to $I \cos \phi$. Hence, the capacitive resistance, X_C, is given by

$$X_C = \frac{\text{supply voltage}}{\text{capacitor current}}$$

$$= \frac{250}{0 \cdot 586}$$

$$= 426 \ \Omega$$

and thus

$$C = \frac{1}{2\pi \times 50 \times 426}$$

$$= 7 \cdot 47 \ \mu\text{F}$$

The reader will probably realise that power factor improvement to unity is in fact making the circuit resonant at this frequency.

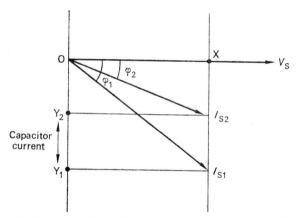

Fig. 2.17   Power factor improvement from $\cos \phi_1$ to $\cos \phi_2$.

Power factor improvement to a value other than unity is shown in Fig. 2.17. By insertion of a shunt capacitor to draw a current equal to $OY_1 - OY_2$ the supply current moves from I_{S1} to I_{S2} and the power factor changes from $\cos \phi_1$ to $\cos \phi_2$. The value of the capacitor is determined as shown above except that, in this case,

$$\text{capacitor current} = I_{S1} \sin \phi_1 - I_{S2} \sin \phi_2$$

If I_{S2} is not known it may be determined as follows:

since
$$I_{S2} \cos \phi_2 = I_{S1} \cos \phi_1$$

$$I_{S2} = I_{S1} \frac{\cos \phi_1}{\cos \phi_2} \tag{2.46}$$

and I_{S1}, $\cos \phi_1$ and $\cos \phi_2$ are known.

Example 2.13
An inductive load connected to a 240 V, 50 Hz supply draws 4·2 A at a power factor of 0·5. Calculate the value of the capacitor required to improve the power factor to 0·9, and determine the apparent power, active power and reactive power at the new power factor.

From eqn. (2.46), since

$$\cos \phi_2 = 0.9$$
$$\cos \phi_1 = 0.5$$

and
$$I_{S1} = 4.2 \text{ A}$$

the new supply current

$$= \frac{4.2 \times 0.5}{0.9}$$

i.e.
$$I_{S2} = 2.33 \text{ A}$$

The capacitor current

$$= I_{S1} \sin \phi_1 - I_{S2} \sin \phi_2$$
$$= 4.2 \times 0.866 - 2.33 \times 0.4347 \text{ (from tables)}$$
$$= 2.62 \text{ A}$$

The capacitive reactance

$$= \frac{\text{supply voltage}}{\text{capacitor current}}$$

$$= \frac{240}{2.62}$$

i.e.
$$X_{\text{C}} = 91.6 \ \Omega$$

and the capacitance from eqn. (2.17)

$$= \frac{1}{2\pi f X_{\text{C}}}$$

hence

$$C = \frac{1}{2\pi \times 50 \times 91 \cdot 6}$$

$$= 34 \cdot 5 \ \mu F$$

The apparent power

$$= \text{voltage} \times \text{current}$$
$$= 240 \times 2 \cdot 33$$
$$= 560 \ VA$$

thus active power

$$= 560 \times \cos \phi_2$$
$$= 560 \times 0 \cdot 9$$
$$= 504 \ W$$

and reactive power

$$= 560 \times \sin \phi_2$$
$$= 560 \times 0 \cdot 4347$$
$$= 243 \ var$$

PROBLEMS ON CHAPTER TWO

(1) When 100 V dc is connected across an inductive coil, the steady-state current is 2·5 A. When 100 V at 50 Hz is connected to the same coil, the r.m.s. value of the current is 2 A. Calculate the coil inductance.

(2) An alternating voltage represented by the expression 150 sin 100 πt is applied to a circuit consisting of a 100 Ω resistor in series with a 0·1 H inductor of resistance 20 Ω. Calculate:

(a) the voltage across the resistor and across the inductor,

(b) the circuit current and its phase displacement from the supply voltage,

(c) the power dissipated in the resistor and the circuit power factor.

(3) Define the term r.m.s. value of an alternating current. What is the average value of a sinusoidally varying current over half a cycle expressed as a fraction of the amplitude? A sinusoidally varying voltage of peak value 100 V is measured by a rectifier instrument calibrated to read r.m.s., consisting of a half-wave rectifier in series with a d'Arsonval movement. The rectifier forward resistance is 50 Ω, the reverse resistance may be assumed infinite. The movement resistance is 10 Ω. Calculate the average value of the instrument current and the scale reading for this voltage.

(4) Describe what is meant by series resonance as applied to an ac circuit. The phase angle of a circuit composed of a 100 Ω resistor

in series with a capacitor is $\pi/3$ rad at 50 Hz. Calculate the value of the inductance required to reduce the phase angle to zero at this frequency.

(5) Describe with the aid of phasor diagrams how the power factor of an inductive circuit may be improved by the use of a capacitor. When a 60 V, 500 Hz supply is connected to an inductive circuit of resistance 50 Ω the supply current is 0·3 A. Calculate the value of the capacitor to reduce the power factor to zero.

(6) Define peak value, average value and r.m.s. value of an alternating current. An alternating current when passed through a resistor immersed in water for 10 min increases the temperature to boiling point. When a direct current of 6 A is passed through the same resistor under identical conditions it takes 16 min to boil the water. Neglecting all other factors other than heat given to the water, calculate the r.m.s. value of the current.

(7) Two circuits connected in parallel across the same alternating voltage take currents of 6 A, phase displacement zero and 10 A phase displacement $\pi/6$ rad respectively. By the use of phasor diagrams calculate the total supply current and its phase displacement from the supply voltage.

(8) Sketch typical impedance/frequency and current/frequency curves for a circuit composed of an inductor in parallel with a capacitor. What is the effect of increasing the coil resistance in such a circuit? The frequencies at which the impedance of such a circuit falls to 0·707 of the maximum value are 750 Hz and 1250 Hz. Calculate the resonant frequency assuming a symmetrical curve and hence the Q factor of the coil. Determine the resistance of a 0·1 mH inductor having this value of Q factor at this frequency.

(9) Define the term 'dynamic impedance' of a parallel tuned circuit. A 1 H, 200 Ω inductor connected in parallel with an 8 μF capacitor is connected to a 20 V variable frequency supply. Calculate:

(a) the frequency at which the supply current falls to a minimum,
(b) the value of the minimum supply current,
(c) the coil current at the frequency at which the supply current is a minimum.

(10) Explain the meaning of the terms 'power factor', 'reactive volt-amperes' and 'power' as applied to a reactive circuit fed with alternating current. A reactive circuit is supplied with a 100 V alternating supply and the resultant current, value 2 A, is displaced from the supply voltage by $\pi/3$ rad. Calculate:

(a) the volt–amperes supplied,
(b) the power dissipated,
(c) the reactive volt-amperes.

(11) Explain why operation of electrical equipment at a low power factor is not desirable. The power factor of a 2000 kVA load at a power factor of 0·8 lagging is to be improved to 0·94 lagging. Describe a method of doing this and determine the kVA required by the additional equipment.

(12) Determine the power taken from a 240 V, 50 Hz supply by a circuit containing three parallel branches drawing currents represented by the expressions $10 \sin \omega t$, $15 \sin [\omega t - (\pi/3)]$ and $15 \sin [\omega t + (\pi/6)]$. Show the phasor diagram.

(13) A series circuit contains a 12 Ω resistor, a 0·0955 H inductor and a capacitor. When a 200 V, 60 Hz supply is connected to the circuit, a current of 10 A flows at a lagging power factor. Find the reactance of the capacitor. Calculate the current and voltages across the components of this circuit when it is connected to a 200 V, 40 Hz supply.

(14) What is meant by the Q factor of a tuned circuit? Explain and describe the effect of Q factor on the following characteristics of such a circuit:

(a) Impedance/frequency,
(b) Current/frequency.

The Q factor of a series tuned circuit is 100 at the resonant frequency of 20 kHz. When 20 V dc is applied to the coil in the circuit, a current of 4 A flows. Calculate:

(a) the inductance of the coil,
(b) the capacitance of the capacitor,
(c) the voltage across the capacitor when a 0·4 V, 20 kHz supply is connected across the circuit.

(15) The Q factor of a series circuit composed of an inductor, capacitor and resistor is 500 at resonance. The voltage across the capacitor when a certain ac supply at the resonant frequency is applied is 5 V and the current is 0·1 A. Calculate:

(a) the supply voltage,
(b) the reactance of the capacitor at resonance,
(c) the resistance in the circuit,
(d) the power dissipated at resonance.

(16) A series tuned circuit having an inductive reactance of 50 Ω and resistance 10 Ω absorbs 100 mW at resonance. Calculate:

(a) the circuit current,
(b) the voltage across inductance and capacitance,
(c) the Q factor,
(d) the supply voltage.

CHAPTER THREE

Network Theorems

3.1 INTRODUCTION

With the increasing complexity of electrical and electronic circuits it becomes essential to find ways of simplifying calculations necessary for design and development. Several useful theorems exist to aid calculations either by providing a simplified approach to the analysis of the existing circuit or by the simplification of the circuit itself. The most commonly applied theorems are

(a) Kirchhoff's laws,
(b) Thévenin's theorem,
(c) Norton's theorem,
(d) Superposition theorem.

These will be dealt with in turn in succeeding sections.

3.2 KIRCHHOFF'S LAWS

The laws are as follows:

(i) the algebraic sum of the currents at any junction is zero,
(ii) the algebraic sum of the voltages around any closed mesh is zero.

The term 'algebraic' in both these laws means that the *direction* of the currents or voltages must be taken into account. The first law put more simply states that the sum of the currents entering a junction is equal to the sum of the currents leaving the junction. The second law states that the sum of the source voltages (electromotive forces) is equal to sum of the voltage drops (potential differences) in any closed mesh. The laws are illustrated in Fig. 3.1.

Figure 3.1(a) shows a junction of four conductors carrying currents I_1, I_2, I_3 and I_4 in the directions shown. Kirchhoff's first law states that the currents entering junction equal the currents leaving junction

$$I_1 + I_4 = I_2 + I_3 \qquad (3.1)$$

Figure 3.1(b) shows a simple one source mesh containing resistors R_1, R_2 and R_3 across which there are voltages V_1, V_2 and V_3.* These voltages act in the same direction, as shown. The source voltage is E_1.

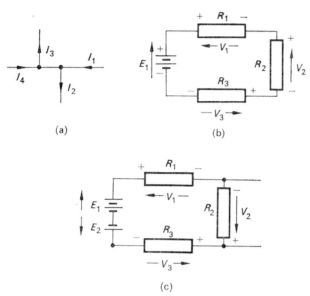

(a)

(b)

(c)

Fig. 3.1 Kirchhoff's laws.

Applying Kirchhoff's second law:

$$E_1 = V_1 + V_2 + V_3 \qquad (3.2)$$

Figure 3.1(c) shows a two source mesh with resistors R_1, R_2, R_3 across which there are voltages V_1, V_2 and V_3. In this circuit the source voltages are acting in opposition and V_2 acts in a direction opposite to V_1 and V_3. (For this to be possible the circuit must be part of a more complex circuit with more than one mesh.) In this case we must take care when applying Kirchhoff's laws to ensure that the polarities are correct within the mesh equation. To do this we take one direction (clockwise or anticlockwise) as positive and apply this in turn to each voltage to ascertain its sign. The actual direction chosen is immaterial provided the choice is not changed partway round the mesh.

* Note that in this and subsequent figures arrows used to depict voltages point to the *positive* side of the voltage.

Applying Kirchhoff's second law

$$-E_2 + E_1 = V_1 - V_2 + V_3 \qquad (3.3)$$

or

$$-E_1 + E_2 = V_2 - V_1 - V_3 \qquad (3.4)$$

Notice that eqn. (3.3) is derived assuming a positive clockwise direction so that E_2 acts to drive a positive clockwise current resulting in potential differences V_1 and V_3. Since E_2 acts in the opposite direction tending to drive an anticlockwise current round the circuit and so producing a p.d. V_2 both E_2 and V_2 are considered negative. Equation (3.4) assumes E_2 and V_2 positive and therefore E_1, V_1 and V_3 negative. Either way the same equation results since eqn. (3.4) is merely eqn. (3.3) with all signs reversed. In more difficult calculations concerning multi-mesh circuits, the problem of ensuring correct signs in the equation very often presents difficulty especially to students who are new to the subject. A simplified approach is suggested below.

(1) Separate the circuit into distinguishable meshes by redrawing if necessary. Number the meshes 1, 2, 3, etc.

(2) In each mesh, sum the source voltages and decide in which direction the resultant current will flow in the mesh. Indicate the resultant current on the mesh numbering i_1, i_2, i_3, etc. If the resultant e.m.f. is zero there will still be a mesh current but any direction may be chosen.

(3) Once all mesh currents are drawn determine the potential differences in each mesh by considering each current in turn and multiplying it by the resistance of the resistors through which it flows. If the current is flowing against the mesh current in the branch common to the two currents, the p.d. caused by it is negative. The p.d. caused by the mesh current is always positive and is equal to the mesh current multiplied by the total series mesh resistance.

(4) Assemble the mesh equations and solve.

The process is demonstrated below.

Example 3.1
Determine the p.d. across the 2 Ω resistor in the circuit of Fig. 3.2(a).

The circuit is redrawn in Fig. 3.2(b); there are three meshes and there will therefore be three simultaneous equations.

In mesh 1 there is no resultant e.m.f. and any direction may be chosen for the mesh current i_1. In mesh 2 the resultant e.m.f. is 2 V acting anticlockwise and mesh current i_2 is drawn in accordingly.

In mesh 3 the e.m.f. is 2 V acting clockwise and mesh current i_3 is drawn in flowing in that direction.

For mesh 1, p.d. due to i_1 = mesh resistance × mesh current
$$= (5 + 4 + 8)i_1$$
$$= 17i_1$$

For the mesh 1 equation this p.d. is positive since it is due to the mesh current i_1.

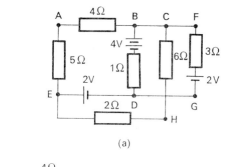

(a)

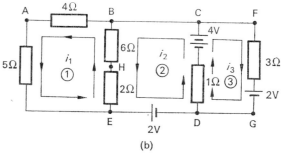

(b)

Fig. 3.2   Relating to Example 3.1.

The p.d. due to i_2 = resistance through which i_2 flows × i_2
$$= -(6 + 2)i_2$$
$$= -8i_2$$

This p.d. is negative in the mesh equation since i_2 flows in the opposite direction to the mesh current i_1 in the branch common to the two currents. There is no p.d. due to i_3; thus we can assemble the mesh 1 equation:

total e.m.f. = total p.d.
$$0 = 17i_1 - 8i_2 \qquad (3.5)$$

For mesh 2, p.d. due to $i_2 = (6 + 2 + 1)i_2$
$$= 9i_2$$

For the mesh 2 equation this p.d. is positive since it is due to the mesh current i_2.

The p.d. due to $i_1 = -8i_1$

This p.d. is negative in the mesh equation since i_1 flows in the opposite direction to the mesh current i_2 in the branch common to the two currents.

The p.d. due to $i_3 = 1i_3$

and is positive in the mesh equation since i_3 flows in the same direction as i_2 in the branch common to the two currents. The mesh equation is

$$\text{total e.m.f.} = \text{total p.d.}$$
$$2 = -8i_1 + 9i_2 + i_3 \tag{3.6}$$

For mesh 3, p.d. due to $i_3 = 4i_3$
$$\text{due to } i_2 = 1i_2$$
$$\text{due to } i_1 = 0$$

Hence, the mesh equation is

$$2 = 1i_2 + 4i_3 \tag{3.7}$$

The three equations are repeated for convenience:

$$0 = 17i_1 - 8i_2 \tag{3.5}$$
$$2 = -8i_1 + 9i_2 + i_3 \tag{3.6}$$
$$2 = i_2 + 4i_3 \tag{3.7}$$

We require the p.d. across the $2\,\Omega$ resistor which is $(i_1 - i_2)2$; hence, we need to find i_1 and i_2 only.
From eqn. (3.5):

$$i_1 = \frac{8}{17}i_2$$

and from eqn. (3.7):

$$i_3 = \frac{2 - i_2}{4}$$

substituting for i_1, i_3 in eqn. (3.6) we have

$$2 = -\frac{64i_2}{17} + 9i_2 + \frac{1}{2} - \frac{i_2}{4}$$

Thus

$$\frac{3}{2} = \left(-\frac{64}{17} + 9 - \frac{1}{4}\right) i_2$$

$$= \frac{(-256 + 612 - 17)}{68} i_2$$

therefore

$$\frac{3}{2} = \frac{339}{68} i_2$$

and therefore

$$i_2 = \frac{204}{678} \text{A} = 0 \cdot 301 \text{ A}$$

from eqn. (3.5)

$$i_1 = \frac{8}{17} i_2$$

$$= \frac{8}{17} \times \frac{204}{678}$$

$$= 0 \cdot 143 \text{ A}$$

$$i_1 - i_2 = 0 \cdot 143 - 0 \cdot 301$$

$$= -0 \cdot 158 \text{ A}$$

This figure represents the current flowing in the $2\,\Omega$ resistor; it is negative which tells us that it is flowing against the direction of the mesh current i_1, *i.e.* in the direction of i_2. The p.d. across the resistor is thus

$$= 0 \cdot 316 \text{ V with H being positive with respect to E}$$

A second example will now be demonstrated without explanation; the reader is advised to work through this example and then attempt the relevant questions at the end of the chapter.

Example 3.2
Calculate the current flowing from the 6 V battery in the circuit shown in Fig. 3.3(*a*).

The circuit is redrawn in Fig. 3.3(*b*) and the mesh currents included in the correct directions.
    The mesh equations are:

for mesh 1 $6 = 10i_1 + 6i_2$ (3.8)

for mesh 2 $8 = 6i_1 + 13i_2 + 4i_3$ (3.9)

for mesh 3 $2 = 4i_2 + 14i_3$ (3.10)

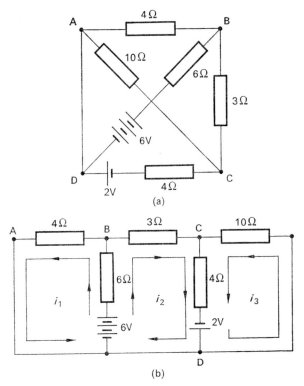

Fig. 3.3   Relating to Example 3.2.

The currents required are i_1 and i_2.
    From eqn. (3.8)

$$i_1 = \frac{6 - 6i_2}{10}$$

From eqn. (3.10)

$$i_3 = \frac{2 - 4i_2}{14}$$

substituting in eqn. (3.9)

$$8 = 6\frac{(6 - 6i_2)}{10} + 13i_2 + 4\frac{(2 - 4i_2)}{14}$$

$$= \frac{18 - 18i_2}{5} + 13i_2 + \frac{4 - 8i_2}{7}$$

$$280 = 126 - 126i_2 + 455i_2 + 20 - 40i_2$$

Thus
$$134 = 289i_2$$
and
$$i_2 = \frac{134}{289}$$
$$i_2 = 0 \cdot 46 \text{ A}$$

from eqn. (3.8)

$$i_1 = (6 - 6 \times 0 \cdot 46) \frac{1}{10}$$
$$= 0 \cdot 322 \text{ A}$$

Current from 6 V battery is $i_1 + i_2 = 0 \cdot 785 \text{ A}$

3.3 THÉVENIN'S THEOREM

This theorem is most useful in that it replaces complex circuits with a simple equivalent circuit consisting of a generator and impedance in series. The theorem states that any circuit composed of active and passive components (*i.e.* voltage or current sources and resistive or reactive components) connected between two terminals, may be replaced exactly by a single circuit composed of a voltage generator in series with an impedance. The generator has a voltage equal to

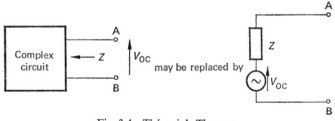

Fig. 3.4 Thévenin's Theorem.

the p.d. between the terminals and the impedance has a value equal to that which exists between the terminals when all sources are replaced by their internal impedances. Thévenin's theorem is equally applicable to ac or dc circuits; at certain frequencies where the effect of the reactive components is negligible one can replace the word 'impedance' in the theorem by 'resistance'. The theorem is illustrated in Fig. 3.4 and is demonstrated in the following examples.

Example 3.3

Calculate the current in the 12 Ω resistor of Fig. 3.5(*a*) using Thévenin's theorem.

The terminals in this case are points A and B, the method being to replace the batteries and 6 Ω and 4 Ω resistor by a series circuit as shown in Fig. 3.5(*b*). The circuit to be replaced is shown in Fig. 3.5(*c*). Notice the 12 Ω load has been removed.

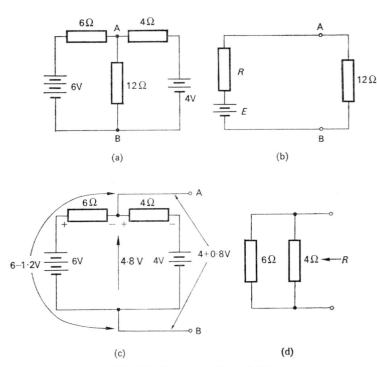

(a)   (b)

(c)   (d)

Fig. 3.5   Relating to Example 3.3.

Since AB is open circuit a circulating current I is set up as shown in Fig. 3.5(*c*). This current is given by

$$I = \frac{\text{mesh e.m.f.}}{\text{mesh resistance}}$$

$$= \frac{6 - 4}{10}$$

$$= 0 \cdot 2 \text{ A}$$

and flows clockwise. The p.d. across the 6 Ω resistor is thus 6 × 0·2, i.e. 1·2 V and has the polarity shown. Taking the mesh left hand branch the p.d. between A and B is thus 6 − 1·2 V, i.e. 4·8 V, A being positive with respect to B. Alternatively, taking the mesh right hand branch, the p.d. across the 4 Ω resistor is 4 × 0·2, i.e. 0·8 V with the polarity shown and the p.d. between A and B is 4 + 0·8 V, i.e. 4·8 V. Notice the p.d. across the 6 Ω resistor is deducted from the 6 V battery e.m.f. in the left hand branch of the mesh but the p.d. across the 4 Ω resistor is added to the 4 V battery e.m.f. in the right hand branch of the mesh. This is due to the respective polarities of the potential differences caused by the clockwise circulating current I being forced by the 6 V battery in the reverse direction through the 4 V battery. (The 4 V battery is actually being charged.)

We now have the e.m.f. E of the equivalent circuit of Fig. 3.5(b) and need to find the resistance R.

To find R the batteries of the circuit in Fig. 3.5(a) are replaced by their internal resistances. Since we are not told what these are we assume that they are zero or, alternatively, included in the 6 Ω and 4 Ω resistances. Thus, we find R by calculating the equivalent resistance of the circuit of Fig. 3.5(d).

$$R = \frac{6 \times 4}{10}$$

$$= 2\cdot4 \,\Omega$$

The current in the 12 Ω resistor in the circuit of Fig. 3.5(b) thus

$$= \frac{4\cdot8}{12 + 2\cdot4}$$

$$= 0\cdot333 \text{ A}$$

which is the required answer.

A second example will now be demonstrated without detailed explanation. Further examples are included at the end of the chapter.

Example 3.4
Calculate the p.d. across XY in the circuit of Fig. 3.6(a) using Thévenin's theorem.

When XY is open circuited the p.d. is equal to the p.d. across the mesh composed of the 6 V and 4 V batteries and the 15 Ω and 5 Ω resistors, i.e. it is equal to the sum of the 6 V battery voltage and the p.d. across the 15 Ω resistor or the 4 V battery voltage and the p.d. across the 5 Ω resistor.

The mesh current

$$= \frac{6 + 4}{15 + 5} \text{ A}$$

$$= 0 \cdot 5 \text{ A}$$

The p.d. across XY is

$$6 - 7 \cdot 5 = -1 \cdot 5 \text{ V}$$

or

$$-4 + 2 \cdot 5 = -1 \cdot 5 \text{ V}$$

with X negative with respect to Y.

The resistance between XY as shown in Fig. 3.6(*b*) is

$$10 + \frac{5 \times 15}{5 + 15}$$

i.e. 13·75 Ω.

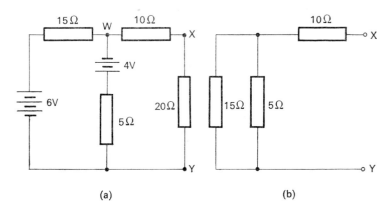

(a)   (b)

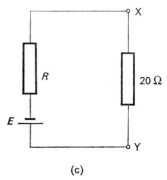

(c)

Fig. 3.6   Relating to Example 3.4.

The equivalent circuit is shown in Fig. 3.6(c) with $E = 1.5$ V and $R = 13.75\ \Omega$.

Hence, current in 20 Ω resistor is

$$\frac{1.5}{33.75}$$

and the p.d. is

$$\frac{20 \times 1.5}{33.75}$$

i.e.

$$\frac{8}{9}\ V$$

with X negative with respect to Y.

3.4 NORTON'S THEOREM

This theorem is the converse of Thévenin's theorem. It has the same advantages in that it replaces a complex circuit by a simple equivalent circuit but in this case the equivalent circuit is composed of a current generator in parallel with an impedance. It is sometimes referred to as the 'parallel generator theorem'.

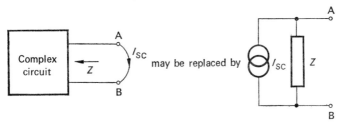

Fig. 3.7 Norton's Theorem.

The theorem states that any circuit made up of active and passive components connected between two terminals may be replaced exactly by an equivalent circuit consisting of a current generator in parallel with an impedance. The current output of the generator is equal to the current which would flow through a short circuit (*i.e.* zero resistance) placed across the terminals. The impedance is equal to the impedance between the terminals when all sources are replaced by their internal impedances. The theorem applies equally at all frequencies and, as with Thévenin's theorem, at certain frequencies only resistances need be considered. The theorem is illustrated in Fig. 3.7 and is demonstrated in the following examples.

Example 3.5
Calculate the current in the 12 Ω resistor of Fig. 3.5(a) using Norton's theorem.

This example is as Example 3.3 and the answer should, of course, be the same. The equivalent circuit is shown in Fig. 3.8(b) the resistance, calculated in Example 3.3 from the circuit of Fig. 3.5(d), is 2·4 Ω.
  The short circuit current is obtained from the circuit of Fig. 3.8(a). Point A is at zero volts with respect to point B; this means that the whole of the 6 V is dropped across the 6 Ω resistor producing a 1 A current flowing towards A and the whole of the 4 V is dropped across the 4 Ω resistor producing 1 A current flowing towards A.

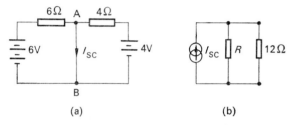

Fig. 3.8   Relating to Example 3.5.

The current flowing away from A, *i.e.* in the short circuit, is thus 2 A by Kirchhoff's first law. The current I_{SC} in the circuit of Fig. 3.8(b) is thus 2 A.
  To determine the current in the 12 Ω resistor we use the principle of current division when two shunt paths are involved. The current in any one path is equal to

$$\frac{\text{resistance of other path}}{\text{sum of path resistances}} \times \text{total current}$$

Thus the 12 Ω resistor current is given by

$$\frac{2 \cdot 4}{12 + 2 \cdot 4} \times 2$$

which is

$$\frac{1}{3} \text{ A}$$

as obtained earlier.
  A second example is now shown and further examples for practice are included at the end of the chapter.

Example 3.6
Calculate the p.d. across XY in the circuit of Fig. 3.6(*a*) using Norton's theorem.

This example is the same as Example 3.4 and the answer should, of course, be the same. As it stands the problem will be more difficult to solve using direct application of Norton's theorem since the presence of the 10 Ω resistor between W and X complicates the finding of the short circuit current. However, since this resistor carries the same current as the 20 Ω resistor we can treat the two as one 30 Ω load to find the current by applying Norton's theorem to the circuit between W and Y instead of to X and Y. We then use this current to find the p.d. across the 20 Ω resistor.

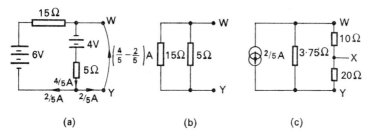

(a)   (b)   (c)

Fig. 3.9   Relating to Example 3.6.

The short circuit current from Fig. 3.9(*a*) is

$$\frac{4}{5} - \frac{6}{15}$$

i.e.

$$\frac{2}{5}\,\text{A}$$

flowing as shown from Y to W.

The shunt resistance of the equivalent circuit of Fig. 3.9(*c*) is obtained from Fig. 3.9(*b*) as

$$\frac{5 \times 15}{5 + 15}$$

i.e. 3·75 Ω.

The current through WY is

$$\frac{3·75}{30 + 3·75} \times \frac{2}{5}\,\text{A}$$

and the p.d. across XY

$$= \frac{3 \cdot 75}{33 \cdot 75} \times \frac{2}{5} \times 20 = \frac{8}{9} \, V$$

as before.

3.5   SUPERPOSITION THEOREM

The superposition theorem is probably the most useful of all those so far considered and is in fact the basis of both Thévenin's and Norton's theorems.

It states that the current flowing in any particular branch of a circuit containing active and passive components is equal to the algebraic sum of the currents due to each individual source when the remaining sources are replaced by their internal impedances. The theorem applies equally to ac and dc circuits, impedances being replaced by resistances at frequencies where the reactive components may be neglected. The theorem is demonstrated for a dc circuit in Fig. 3.10, the current I in part (a) being equal to the current i_1 in part (b) added algebraically to current i_2 in part (c). In Fig. 3.10(b), E_2 is replaced by its internal impedance R_2; in Fig. 3.10(c), E_1 is replaced by its internal impedance R_1. If R_1 and R_2 are zero then the batteries are replaced by short circuits.

The theorem is demonstrated in the following examples.

Example 3.7
Calculate the current in the 12 Ω resistor of Fig. 3.5(a) using the superposition theorem.

Replacing the 4 V battery by a short circuit as in Fig. 3.11(a) we have the circuit resistance as

$$6 + \frac{48}{16} = 9 \, \Omega$$

and the battery current

$$= \frac{6}{9}$$

$$= \frac{2}{3} \, A$$

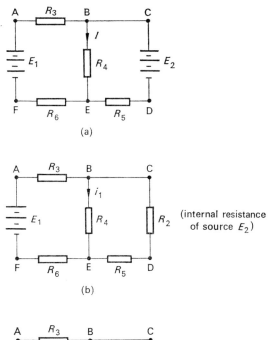

(a)

(b)

(c)

Fig. 3.10 Superposition Theorem.

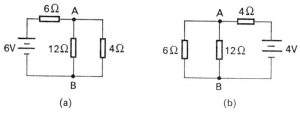

(a)   (b)

Fig. 3.11 Relating to Example 3.7.

The current flowing in the 12 Ω resistor (using the current distribution law)

$$= \frac{4}{16} \times \frac{2}{3}$$

$$= \frac{1}{6} \text{ A}$$

from A to B.

In the circuit shown in Fig. 3.11(b), the 6 V battery is replaced by a short circuit.

Circuit resistance

$$= 4 + \frac{72}{18}$$

$$= 8 \ \Omega$$

The battery current

$$= \frac{4}{8}$$

$$= \frac{1}{2} \text{ A}$$

The current flowing in the 12 Ω resistor

$$= \frac{6}{18} \times \frac{1}{2}$$

$$= \frac{1}{6} \text{ A}$$

from A to B.

Thus, total current

$$= \frac{1}{3} \text{ A}$$

as before.

Example 3.8

Calculate the p.d. across XY in the circuit of Fig. 3.6(a) by the superposition theorem.

The two separate circuits are shown in Figs. 3.12(a) and (b). In Fig. 3.12(a):

circuit resistance

$$= 15 + \frac{150}{35}$$

$$= \frac{135}{7}\,\Omega$$

6 V battery current

$$= \frac{6}{135/7}$$

$$= \frac{42}{135}\,A$$

Current in XY branch

$$= \frac{5}{35} \times \frac{42}{135}$$

$$= \frac{6}{135}\,A$$

flowing from X to Y.

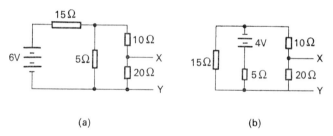

(a)   (b)

Fig. 3.12   Relating to Example 3.8.

In Fig. 3.12(b):

circuit resistance

$$= 5 + \frac{30 \times 15}{45}$$

$$= 15\,\Omega$$

4 V battery current

$$= \frac{4}{15}\,A$$

Current in XY branch

$$= \frac{15}{45} \times \frac{4}{15}$$

$$= \frac{4}{45} \text{ A}$$

from Y to X

Thus resultant current in XY is

$$\frac{4}{45} - \frac{6}{135}$$

$$= \frac{2}{45} \text{ A}$$

from Y to X.

Therefore the p.d. across the 20 Ω resistor

$$= \frac{2}{45} \times 20$$

$$= \frac{8}{9} \text{ V}$$

with X negative with respect to Y.

Further examples are included at the end of the chapter.

3.6   INPUT AND OUTPUT IMPEDANCES; MATCHING

The input impedance of any circuit however complex is defined as the ratio of the input voltage to the input current. The output impedance is the impedance 'seen' looking back into the output terminals with all active sources replaced by their internal impedances. It follows that the impedance determined by Thévenin's or Norton's theorem is, in fact, the output impedance.

The concept of input and output impedances is most useful in that with the aid of the theorems described above it enables us to replace what may be a complex circuit by a simple circuit composed of a generator and two impedances.

In certain applications, especially electronics, we are concerned with transfer of a signal from one circuit to another. The signal may be voltage, current or power as described in Chapter 7. If the two circuits are replaced by a generator and impedance (the output impedance of the driver circuit), and by a load impedance (the input impedance of the circuit to be supplied), it follows from the Thévenin equivalent circuit that, for maximum voltage transfer from one circuit to the next, the ratio of load impedance to output impedance should be high; from the Norton equivalent circuit it follows that for maximum current transfer the ratio of load impedance to output impedance should be low. Since power is the product of voltage and current, it is logical, and is in fact shown below, that for maximum power transfer from the driver circuit to the load circuit, the output and input impedances should be equal and thus their ratio is unity.

Of course, the ratio of input to output impedance is not necessarily correct for the particular purpose, and in this case it is necessary to introduce an additional circuit or piece of equipment such as a transformer (see Chapter 5) so that the input impedance of the load circuit appears to the driver circuit to be of the correct value relative to the driver output impedance. This process of adjusting the impedance ratio is called *matching*.

3.7 THE MAXIMUM POWER TRANSFER THEOREM

Consider two circuits coupled so that one drives the next, *i.e.* supplies power to the next. The simple equivalent circuit using Thévenin's theorem consists of a voltage source and two resistors in series. One resistor is the source resistance, *i.e.* the output resistance of the driver circuit, the other is the input resistance of the driven circuit and is referred to as the load (*see* Fig. 3.13).

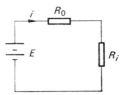

Fig. 3.13   Maximum power transfer theorem.

In the circuit of Fig. 3.13, the current from one circuit to the next is given by

$$i = \frac{E}{R_o + R_i}$$

where R_o is the output resistance and R_i is the input resistance.

The power in the load resistor R_i equals $i^2 R_i$ and is given by

$$\text{power} = \left(\frac{E}{R_i + R_o}\right)^2 R_i$$

$$= \frac{E^2 R_i}{R_i{}^2 + 2R_i R_o + R_o{}^2}$$

$$= \frac{E^2 R_i}{(R_i - R_o)^2 + 4R_i R_o} \tag{3.11}$$

For the right hand side of eqn. (3.11) to be a maximum the denominator should be a minimum. By examination it can be seen that this occurs when $(R_i - R_o)$ is zero, $i.e.$ when $R_i = R_o$. For all other values of R_i and R_o, the term $(R_i - R_o)^2$ is positive and increases the denominator value above $4R_i R_o$, the value of the denominator when $R_i = R_o$.

Maximum power transferred when $R_i = R_o$

$$= \frac{E^2 R_i}{4R_i R_o}$$

$$= \frac{E^2}{4R_o} \tag{3.12}$$

Example 3.9
A circuit consists of a 4 V battery of negligible internal resistance, in series with two resistors of value 6 Ω and 2 Ω respectively. A variable resistor is connected across the 2 Ω resistor. Determine the value of the variable resistor when it is dissipating maximum power and determine the value of the maximum power dissipated.

The voltage across the 2 Ω resistor when the load (variable) resistor is not connected is $(2/8) \times 4$, $i.e.$ 1 V. This is the Thévenin equivalent circuit supply voltage.

The resistance of the supply circuit looking back into the circuit across the 2 Ω resistor is (2 in parallel with 6) ohms, $i.e.$ 3/2 Ω. This is the output resistance of the circuit and the source resistance of the Thévenin equivalent circuit. Thus maximum power is transferred when the variable resistor is equal to 3/2 Ω.

From eqn. (3.12) the maximum power

$$= \frac{E^2}{4R_o}$$

where E (the Thévenin equivalent circuit supply voltage) is 1 V, and R_o (the output resistance) is 3/2 Ω.

Thus, maximum power

$$= \frac{2}{4 \times 3}$$

$$= \frac{1}{6} \text{ W}$$

PROBLEMS ON CHAPTER THREE

(1) State Kirchhoff's laws as applied to an electrically conductive circuit. A 4 V battery of internal resistance 3 Ω is connected in parallel with a 2 V battery of resistance 1 Ω and the combination then applied to a 5 Ω resistor. Using Kirchhoff's laws determine the voltage across the 5 Ω resistor.

(2) State Thévenin's theorem and briefly discuss its usefulness in circuit analysis. A source consisting of two batteries in parallel, one 4 V, 3 Ω, the other 2 V, 1 Ω supplies current to a 10 Ω load. Calculate the value of the current using Thévenin's theorem.

(3) An 8 V battery with internal resistance 4 Ω is connected across a 12 Ω load. A 0–5 Ω variable resistor is then connected across the 12 Ω load and varied continuously from zero to maximum value. Calculate the minimum wattage rating of the variable resistor in order that it should not be damaged during the varying of its resistance.

(4) Describe how the superposition theorem may be used in circuit analysis. A source composed of a 4 V, 3 Ω battery in parallel with a 2 V, 1 Ω battery is connected to a 5 Ω load. Use the superposition theorem to determine the load current.

(5) State the maximum power transfer theorem and explain briefly what is meant by matching between circuits. A certain circuit consists of a voltage source connected in series with a 10 Ω resistor and a resistor of resistance R Ω. A variable resistor is now connected across R and adjusted to 6 Ω when maximum power of 4 W is obtained in the variable resistor. Calculate:

(a) the value of R,
(b) the value of the voltage source.

(6) State Norton's theorem and illustrate the simple equivalent circuit of any network obtained by this theorem. A π network consisting of three 6 Ω resistors is inserted between a 2 V source of zero internal resistance and a 6 Ω load. Use Norton's theorem to determine the load current and hence the load voltage.

(7) A 1 V, 3 Ω source is connected to a load circuit consisting of a 6 Ω resistor in parallel with a 12 Ω resistor. Determine the maximum

power available from this source, the power dissipated in the load and hence the percentage of the maximum power which is transferred from source to load under these conditions.

(8) A certain resistive network has four terminals A, B, C and D. The following resistors are connected between the terminals:

$$AB\ 6\ \Omega,\ BC\ 4\ \Omega,\ CD\ 5\ \Omega,\ DA\ 5\ \Omega,\ BD\ 8\ \Omega$$

Determine the current in BD when a 3 V battery of zero internal resistance is connected across AC.

(9) A certain electronic amplifier circuit may be considered to be made up of two stages, one driving the other. Each stage has an input resistance of 40 Ω and the output may be represented by a Thévenin equivalent circuit containing a resistance of 1000 Ω and a voltage generator whose output is ten times the input voltage applied to the stage. The two-stage circuit is loaded correctly for maximum power transfer and the output power is 0·1 W. Determine the voltage at the input on the first stage.

(10) The open circuit voltage of a battery is 9 V. When connected to a 20 Ω load the battery voltage is reduced to 8·5 V. Determine the load current when three of these batteries connected in parallel supply a 50 Ω load.

CHAPTER FOUR

Three-phase Systems

4.1 INTRODUCTION

The generation, transmission and distribution of alternating current throughout the United Kingdom is a *multiphase* system, in that three voltages are generated simultaneously. These voltages are displaced from one another in phase by 120 electrical degrees $(2\pi/3$ rad) and are distinguished from one another either by a number system (1, 2 and 3) or a colour system (red, yellow and blue). Since

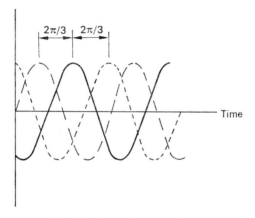

Fig. 4.1 Three-phase waveform.

they are generated simultaneously in one machine they alternate at the same frequency (50 Hz). The alternators are so wound that the amplitudes off load are equal (*see* Fig. 4.1).

The reasons for using a multi- rather than a single-phase system are numerous. They include:

(*a*) more efficient generation, transmission and distribution,

99

(b) three-phase machines have a more even torque, use less copper per unit power, have a higher efficiency and power factor and are self starting,

(c) industrial and domestic supplies are available from a single source.

The methods of interconnection of the three phase windings both in the generator and load are considered in the following sections.

4.2   METHODS OF CONNECTION

The three single-phase supplies generated as described above may be transmitted using six conductors, two per phase. However, this arrangement is expensive and, as it turns out, is not necessary because of the phase displacement between voltages. There are two methods of interconnecting the windings at the alternator (and thus the loads and also windings of transformers encountered along the distribution path). These are *delta* or mesh connection and *star* connection. The arrangements are shown in Fig. 4.2.

In the delta connection, shown in Fig. 4.2(a) the windings are connected in a continuous closed loop in such a way that at the instant when any particular phase voltage is assumed positive, its direction of action is along the arrow shown, *e.g.* when the red-phase voltage V_R is positive, point G is positive with respect to point J in the diagram. It should be noted that when V_R is positive, V_B and V_Y may be positive or negative depending upon the instant chosen along the time axis of Fig. 4.1. The point of importance is that when V_B or V_Y are positive they act in the direction shown, *i.e.* H positive with respect to G or J positive with respect to H, respectively. For windings connected in this manner the resultant voltage across the three coils, *i.e.* between J and J' in Fig. 4.2(a) is zero. This can be shown using eqns. (2.8) and (2.9) as follows.

If the equation describing V_R is

$$V_R = V_{R(max)} \sin \omega t \tag{4.1}$$

then since V_Y lags V_R by $2\pi/3$ rad and V_B lags V_Y by $2\pi/3$ rad

$$V_Y = V_{Y(max)} \sin \left(\omega t - \frac{2\pi}{3} \right) \tag{4.2}$$

and

$$V_B = V_{B(max)} \sin \left(\omega t - \frac{4\pi}{3} \right) \tag{4.3}$$

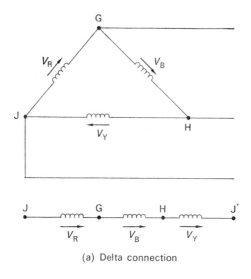

(a) Delta connection

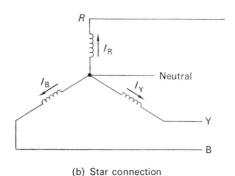

(b) Star connection

Fig. 4.2 Delta and star connections.

and from Fig. 4.2(a) voltage between J and J$'$ = $V_R + V_Y + V_B$
= $V_{R(max)} \sin \omega t + V_{Y(max)} \sin (\omega t - 2\pi/3) + V_{B(max)} \sin (\omega t - 4\pi/3)$
which, if

$$V_{R(max)} = V_{Y(max)} = V_{B(max)}$$

is equal to zero. Hence since the resultant voltage between J and J$'$ is zero, point J can be joined to J$'$ as in Fig. 4.2(a). The condition that the maximum values of voltage are equal applies to the generator off load, if the windings have equal turns and are similarly constructed, but not necessarily to the load. This is considered later.

The advantage of this connection is that three wires out of the six are no longer necessary, with consequent gain economically but without loss, since the three voltages are still available. The three wires connected to points G, H and J are called *lines* and the voltages between them and the currents in them are called line voltages and currents to distinguish them from phase voltages and currents. From the diagram it is clear that the voltage between any two lines is equal to the phase voltage of the winding between the lines. However, the line currents do not equal the phase currents since the current in any line is the phasor sum of the two phase currents in the windings connected to the line.

The *star* connection is shown in Fig. 4.2(b). In this connection the windings are connected so that the direction of the voltages when assumed positive is outwards. Provided the windings are similarly constructed and have equal turns, the current in the fourth wire, called the *neutral* wire, is zero. This can be shown using eqns. (2.7) and (2.9) as follows: if the equation describing the current in the 'red' phase, i_R, is

$$i_R = I_{R(max)} \sin \omega t \tag{4.4}$$

then since i_Y lags i_R by $2\pi/3$ rad, the equation for i_Y is

$$i_Y = I_{Y(max)} \sin\left(\omega t - \frac{2\pi}{3}\right) \tag{4.5}$$

and similarly

$$i_B = I_{B(max)} \sin\left(\omega t - \frac{4\pi}{3}\right) \tag{4.6}$$

where i_Y and i_B are the instantaneous values of the 'yellow' and 'blue' phases respectively.

The current in the fourth wire at any instant

$$i_N = i_R + i_Y + i_B$$

$$= I_{R(max)} \sin \omega t + I_{Y(max)} \sin\left(\omega t - \frac{2\pi}{3}\right) + I_{B(max)} \sin\left(\omega t - \frac{4\pi}{3}\right)$$

which if expanded yields zero provided that

$$I_{R(max)} = I_{Y(max)} = I_{B(max)}$$

The neutral wire can, in fact, be dispensed with under these conditions.

Both methods of connection are further examined in more detail below.

4.3   DELTA CONNECTED SYSTEMS

As was stated in Section 4.2, it is clear that the voltage between any two lines of a delta-connected system is equal to the voltage across the phase winding connected between the lines. The relationship between line and phase currents will now be deduced. In the delta connection the windings are joined so that the voltages act round the mesh when positive; it follows that the 'positive' direction of currents is also round the mesh as shown in Fig. 4.3. In this figure the currents in the red, yellow and blue phases are denoted I_R, I_Y and I_B respectively. It can be seen that the current in the line connected to G is the phasor difference of I_R and I_B since one current flows towards G when positive and the other flows from G. The current in line G, I_G, is thus given by

$$I_G = I_R - I_B \qquad (4.7)$$

where the right hand side denotes the phasor difference, *i.e.* taking into consideration direction of flow. It should be realised, of course, that at certain instances in time both currents i_R and i_B flow

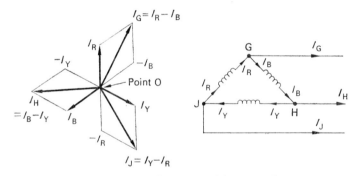

Fig. 4.3   Phasor diagram for delta connection.

towards G. However, eqn. (4.7) still applies since at these instances the red phase current is still denoted by i_R but the blue phase current is negative (the positive direction is from G to H) and is denoted by $-i_B$ so that eqn. (4.7) for instantaneous values would read

$$i_G = i_R - (-i_B)$$
$$= i_R + i_B$$

which indicates that the instantaneous value of the G line current equals the sum of the instantaneous values of i_R and i_B at these moments in time.

The rules for obtaining phasor differences of two time varying quantities are the same as those for obtaining vector differences, *i.e.* the quantity to be subtracted is drawn in the opposite direction to that used for positive and the resultant is then obtained using the usual parallelogram method. The phasor for the line current I_G is thus as shown in Fig. 4.3 being the sum of I_R and $-I_B$. Similarly the line currents I_H and I_J, given by equations

$$I_H = I_B - I_Y \tag{4.8}$$

$$I_J = I_Y - I_R \tag{4.9}$$

have phasors as shown in the diagram.

If I_R, I_Y and I_B have equal amplitudes and are displaced by $2\pi/3$ rad from one another it is clear that the line currents I_G, I_H, I_J also have equal amplitudes and are similarly displaced from one another by $2\pi/3$ rad. Examination of the diagram shows that I_R and $-I_B$ are displaced from one another by $\pi/3$ rad so that

$$I_G = I_R \cos \frac{\pi}{6} + I_B \cos \frac{\pi}{6}$$

$$= (3)^{\frac{1}{2}} I_R \quad \text{or} \quad (3)^{\frac{1}{2}} I_B \quad \text{if} \quad I_R = I_B \tag{4.10}$$

Similarly

$$I_H = (3)^{\frac{1}{2}} I_B \quad \text{or} \quad (3)^{\frac{1}{2}} I_Y \tag{4.11}$$

and

$$I_J = (3)^{\frac{1}{2}} I_R \quad \text{or} \quad (3)^{\frac{1}{2}} I_Y \tag{4.12}$$

or, in general,

$$\text{line current} = (3)^{\frac{1}{2}} \times \text{phase current} \tag{4.13}$$

provided that the system has equal amplitude currents equally displaced, *i.e.* is a balanced system. Note that in the discussion of phasor diagrams the symbols I_R, I_Y, I_B, etc. may denote either amplitude or r.m.s. values.

Summarising, for a delta connected balanced system,

$$\text{line voltage} = \text{phase voltage} \tag{4.14}$$

$$\text{line current} = (3)^{\frac{1}{2}} \times \text{phase current} \tag{4.13}$$

4.4   STAR CONNECTED SYSTEMS

For a star connected system the line current is equal to the current in the phase winding to which the line is connected. It was stated earlier that in a star system the voltages act in a direction outwards

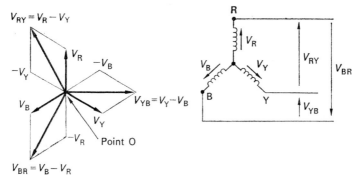

Fig. 4.4 Phasor diagram for star connection.

from the common (neutral) point, when they are positive. It follows from Fig. 4.4, therefore, that the voltage between lines is the phasor difference of the voltages across the two phase windings. (The point concerning instantaneous directions applies equally as before.) So that, from the diagram:

$$\text{voltage between red–yellow lines } V_{RY} = V_R - V_Y \quad (4.15)$$

$$\text{voltage between yellow–blue lines } V_{YB} = V_Y - V_B \quad (4.16)$$

$$\text{voltage between blue–red lines } V_{BR} = V_B - V_R \quad (4.17)$$

The phasor diagram shown in Fig. 4.4 indicates that there is a $\pi/3$ rad displacement between V_R and $-V_Y$, between V_Y and $-V_B$ and between V_B and $-V_R$ so that

$$V_{RY} = V_R \cos\frac{\pi}{6} + V_Y \cos\frac{\pi}{6}$$

$$= (3)^{\frac{1}{2}} V_R \quad \text{or} \quad (3)^{\frac{1}{2}} V_Y \quad (4.18)$$

if

$$V_R = V_Y$$

Similarly

$$V_{YB} = (3)^{\frac{1}{2}} V_Y \quad \text{or} \quad (3)^{\frac{1}{2}} V_B \quad (4.19)$$

and

$$V_{BR} = (3)^{\frac{1}{2}} V_B \quad \text{or} \quad (3)^{\frac{1}{2}} V_R \quad (4.20)$$

if

$$V_B = V_R = V_Y$$

or in general,

$$\text{voltage between lines} = (3)^{\frac{1}{2}} \text{ phase voltage} \quad (4.21)$$

Summarising, for a balanced star connected system

$$\text{line current} = \text{phase current} \qquad (4.22)$$

$$\text{line–line voltage} = (3)^{\frac{1}{2}} \times \text{phase voltage} \qquad (4.21)$$

$$\text{line–neutral voltage} = \text{phase voltage} \qquad (4.23)$$

It is common practice to use a star connection with neutral on the distribution side of the supply network so that single-phase and three-phase supplies are available from the same system (single-phase between line and neutral). This can, of course, lead to unbalance in the system and special analyses must be made to determine line or phase currents and voltages.

4.5 POWER IN A BALANCED THREE-PHASE LOAD

Consider a three-phase supply loaded on each phase such that the phase currents and voltages are of equal amplitude and equally displaced from one another in phase. If the r.m.s. values of line voltage and current are denoted by V_L and I_L, the r.m.s. values of phase voltage and current are denoted by V_{ph} and I_{ph}, and the power factor in each phase load is denoted by $\cos \varphi$, then

$$\text{power per phase} = V_{ph}I_{ph} \cos \varphi \qquad (4.24)$$

and

$$\text{total power} = 3V_{ph}I_{ph} \cos \varphi \qquad (4.25)$$

For a delta system,

$$V_L = V_{ph}$$
$$I_L = (3)^{\frac{1}{2}}I_{ph}$$

so that

$$\text{total power} = 3\frac{V_L I_L}{(3)^{\frac{1}{2}}} \cos \varphi$$

$$= (3)^{\frac{1}{2}}V_L I_L \cos \varphi \qquad (4.26)$$

For a star system,

$$V_L = (3)^{\frac{1}{2}}V_{ph}$$
$$I_L = I_{ph}$$

so that

$$\text{total power} = 3\frac{V_L I_L}{(3)^{\frac{1}{2}}} \cos \varphi$$

$$= (3)^{\frac{1}{2}}V_L I_L \cos \varphi \qquad (4.26)$$

i.e. the expression for total power is the same, regardless of how the system is connected; provided that the system is balanced.

Example 4.1
Three 10 μF capacitors are star connected across a 400 V, 50 Hz three-phase supply. Find the current per phase.

The reactance in each phase

$$= \frac{10^6}{2\pi \times 50 \times 10} \text{ ohms}$$

$$= 318 \cdot 2 \text{ ohms}$$

The line voltage is 400 V and since the system is star connected, the phase voltage

$$= \frac{400}{(3)^{\frac{1}{2}}} \text{ volts}$$

Thus, the phase current

$$= \frac{400}{(3)^{\frac{1}{2}} \times 318 \cdot 2} \text{ amperes}$$

$$= 0 \cdot 725 \text{ A}$$

The system is balanced, therefore this current flows in each phase.

Example 4.2
Three identical resistors are connected in star across a 440 V, three-phase supply. Determine the value of each resistor and the total power consumed if the line current is 4 A. If the resistors are now connected in delta across the same supply, what would be the line current and total power?

The line voltage is 440 V.
    The phase voltage is $440/(3)^{\frac{1}{2}}$, *i.e.* 254 V since the loads are star connected.
    The phase current, which equals the line current, is 4 A.
    The load resistance per phase is, therefore, 254/4 Ω, *i.e.* 63·5 Ω.
    Total power, from eqn. (4.26)

$$= (3)^{\frac{1}{2}} \times 440 \times 4$$

since the power factor is unity,

$$\text{total power} = 3050 \text{ W}$$

If the resistors are connected in delta, the phase and line voltages are the same and equal 440 V,

$$\text{phase current} = \frac{440}{63\cdot5}$$

$$= 6\cdot92 \text{ A}$$

$$\text{line current} = (3)^{\frac{1}{2}} \times 6\cdot92 \text{ A}$$

And the total power

from eqn. (4.26),

$$= (3)^{\frac{1}{2}} \times 440 \times (3)^{\frac{1}{2}} \times 6\cdot92$$

$$= 9150 \text{ W}$$

Example 4.3
A three-phase delta connected motor operating from a 400 V, 50 Hz supply has an output of 40 bhp, the efficiency and power factor being 95 per cent and 0·9 respectively. Calculate the current in each phase of the stator winding.

The output of the motor

$$= 40 \text{ bhp}$$

$$= 40 \times 746 \text{ W}$$

Since the efficiency is 95 per cent, the input to the motor

$$= \frac{40 \times 746}{0\cdot95} \text{ W}$$

$$= 31\,400 \text{ W}$$

Total input power

from eqn. (4.26) where

$$= (3)^{\frac{1}{2}} V_{\text{L}} I_{\text{L}} \cos \varphi$$

$$V_{\text{L}} = 400 \text{ V}$$

I_{L} is required

$$\cos \varphi = 0\cdot8$$

So,

$$\text{input} = (3)^{\frac{1}{2}} \times 400 \times I_{\text{L}} \times 0\cdot8$$

and this

$$= 31\,400 \text{ from above}$$

Thus,

$$I_{\text{L}} = \frac{31\,400}{(3)^{\frac{1}{2}} \times 400 \times 0\cdot8}$$

$$= 64 \text{ A}$$

The motor is delta connected so the phase current is equal to line current divided by 3.

Thus phase current

$$= \frac{64}{(3)^{\frac{1}{2}}}$$

$$= 36 \cdot 9 \text{ A}$$

Assuming balanced conditions this is the current in each phase winding of the stator.

Example 4.4

A three-phase 500 V star connected ac generator supplies a three-phase 100 hp mesh connected induction motor, the efficiency being 90 per cent and the power factor 0·85. Find the current (a) in each motor phase, (b) in each generator phase.

$$\text{motor output} = 100 \text{ hp}$$
$$= 74\,600 \text{ W}$$

Since the efficiency is 90 per cent the motor input

$$= \frac{74\,600}{0 \cdot 9} \text{ W}$$

$$= 82\,888 \text{ W}$$

The input power

$$= (3)^{\frac{1}{2}} V_L I_L \cos \varphi$$

from eqn. (4.26), so that input power

$$= (3)^{\frac{1}{2}} \times 500 \times I_L \times 0 \cdot 85 \text{ W}$$

which, from above,

$$= 82\,888 \text{ W}$$

Hence,

$$I_L = \frac{82\,888}{(3)^{\frac{1}{2}} \times 500 \times 0 \cdot 85}$$

$$= 112 \cdot 6 \text{ A}$$

This is in fact the current flowing in the lines between generator and motor. Since the generator is star connected this is also the generator phase current. The motor is mesh (delta) connected, so

$$\text{motor phase current} = \frac{112 \cdot 6}{(3)^{\frac{1}{2}}}$$

$$= 65 \text{ A}$$

PROBLEMS ON CHAPTER FOUR

(1) Show by phasor diagrams the relationship between line currents and phase currents in a balanced delta connected three-phase system. A balanced delta connected load takes 50 kW at 0·75 power factor from a three-phase 440 V supply. Calculate the line current.

(2) Three equal impedances are connected in star to a 440 V, 50 Hz supply. The power factor is 0·9 lagging and 15 kVA is drawn from the supply. Determine the line current and total power drawn from the same supply if the same impedances were delta connected.

(3) A three-phase load consists of three equal impedances, each having a resistive component of 10 Ω and a capacitive reactive component of 15 Ω. The supply voltage is 400 V, 50 Hz. Calculate line and phase currents if the load is (a) delta connected, (b) star connected.

(4) Discuss the advantages of three-phase working compared with single-phase working. A three-phase star connected alternator provides 240 V per phase to a delta connected load composed of equal impedances of resistance 5 Ω and inductive reactance 10 Ω. Calculate:

(a) the load phase current,
(b) the load power factor,
(c) the alternator current,
(d) the alternator output in watts.

(5) Describe the two methods of connection used in three-phase ac systems. Calculate the total power supplied to a star connected resistive load of 100 Ω per phase when the line supply voltage is 200 V.

(6) A 440 V three-phase star connected motor has an output of 60 bhp with an efficiency of 92 per cent and a power factor of 0·9. Calculate the line current and draw the phasor diagram showing all voltages and currents in the motor.

(7) A three phase, 25 hp, 440 V, 50 Hz motor has a full-load efficiency of 90 per cent and a full-load power factor of 0·75 lagging. Calculate the input kW, kVA, and kvar. Determine the full-load phase currents if the motor is delta connected.

(8) A delta connected load consisting of three equal impedances of 25 Ω resistance and 20 Ω reactance is supplied with 440 V line voltage by a three-phase star connected alternator. Calculate:

(a) the line current between alternator and load,
(b) the alternator output in kW and in kVA.

Neglect the line losses.

CHAPTER FIVE

Transformers

5.1 FUNDAMENTALS OF TRANSFORMER ACTION

Transformer action depends upon electromagnetic induction as described by Michael Faraday's law discussed in Section 1.3. It was stated there that whenever a magnetic field surrounding or in the vicinity of a conductor changes in magnitude (*i.e.* the flux density changes), a voltage is induced across the conductor. The voltage will act in a direction so as to oppose the change which causes it. This aspect of opposition is discussed in more detail in Chapter 6 when fundamentals of rotating machinery are discussed.

Basically a transformer consists of two or more coils linked by a common magnetic circuit of low reluctance. An alternating supply applied to one coil, called the primary winding, sets up an alternating current and thus a changing magnetic field. This field is linked to the other coil, called the secondary winding so that a voltage is induced across it. The induced voltage may then be used as a supply. The principal advantage, which is further discussed below, is that the magnitude of the secondary voltage may be adjusted so that it is possible to obtain any level of voltage from a single supply. Transformers find wide application in distribution where the generated voltage is first increased to improve transmission efficiency and then reduced progressively to a safe working level for the consumer, in electronic equipment power supplies and in electronic amplifiers, where the matching ability of a transformer is put to use. Matching was discussed briefly in Chapter 3.

5.2 TRANSFORMER CONSTRUCTION

Transformers for use at power and audio frequencies (to about 20 kHz) use metal cores for the magnetic circuit. At frequencies above this range dust cores and even air cores may be used because of the increasing losses at high frequencies.

Metal cores are made of high permeability steel alloys formed into thin laminations (about 0·35 mm thick) which are bolted together. Each lamination is insulated from its neighbour. The purpose of a laminated rather than solid core is to reduce the effect of small circulating currents, called eddy currents, which are induced within the core by the changing flux. The core shape may be either of two shown in Fig. 5.1. Part (*a*) is generally known as 'core construction', and part (*b*) is known as 'shell construction'. The majority of power transformer cores use the configuration of Fig. 5.1(*a*).

Transformer windings are generally circular in cross-section to withstand the considerable mechanical stresses set up when on load. There are various types of winding depending upon the intended use of the transformer.

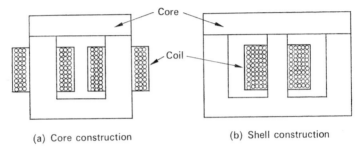

(a) Core construction   (b) Shell construction

Fig. 5.1   Transformer construction.

Once assembled, the windings and core are contained within a metal case or tank. Cooling is effected either by air or, on larger power transformers, by oil, and provision is made for the coolant to circulate through the container.

5.3 THE IDEAL TRANSFORMER

If certain assumptions are made and the transformer is considered ideal, various basic relationships between voltages and currents at the primary and secondary sides of the transformer may be derived. In practice these relationships are approximate but nevertheless useful. To idealise the transformer the following assumptions are made:

(*a*) zero winding resistance,
(*b*) no flux losses, *i.e.* total flux links primary and secondary,
(*c*) zero core losses,

(*d*) the core permeability is extremely high so a negligible m.m.f. is required to set up the flux.

Consider the circuit of Fig. 5.2 which shows an ideal transformer connected between a supply and a load. Provided that the flux has the same waveform as the current which is assumed sinusoidal, the equation describing the flux may be written

$$\Phi = \Phi_{max} \sin 2\pi ft \qquad (5.1)$$

where Φ is the instantaneous value of flux (Wb)
  Φ_{max} is the maximum value of flux (Wb)
  f is frequency (Hz)
  t is time (s).

Kirchhoff's voltage law applied to the primary circuit indicates that the voltage induced in the primary winding by the changing flux E_p, is equal to and opposes the applied voltage V_p.

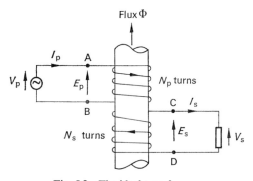

Fig. 5.2   The ideal transformer.

According to Faraday's law this induced voltage is equal to number of turns times the rate of change of flux. Differentiating eqn. (5.1) gives

rate of change of flux with time $= 2\pi f\Phi_{max} \cos 2\pi ft \qquad (5.2)$

so that

$$E_p = N_p \times 2\pi f\Phi_{max} \cos 2\pi ft \qquad (5.3)$$

which is a cosinusoidal wave, *i.e.* a sinusoidal wave displaced from the flux wave by $\pi/2$ rad leading, with an amplitude $2\pi fN_p\Phi_{max}$.

The r.m.s. value of this voltage

$$= \frac{\text{peak value}}{(2)^{\frac{1}{2}}}$$

thus

$$E_p = \frac{2\pi}{(2)^{\frac{1}{2}}} f N_p \Phi_{max}$$

$$= 4{\cdot}44 f N_p \Phi_{max} \tag{5.4}$$

and since

$$E_p = V_p$$

$$V_p = 4{\cdot}44 f N_p \Phi_{max} \tag{5.5}$$

where E_p and V_p are the r.m.s. values of primary induced voltage and applied voltage respectively. It follows that

$$\Phi_{max} = \frac{V_p}{4{\cdot}44 f N_p} \tag{5.6}$$

The changing flux links the secondary winding and the secondary voltage E_s is induced. From Faraday's law and eqn. (5.4) it follows that

$$E_s = 4{\cdot}44 f N_s \Phi_{max} \tag{5.7}$$

Neglecting winding impedances

$$E_s = V_s$$

so that

$$V_s = 4{\cdot}44 f N_s \Phi_{max} \tag{5.8}$$

and substituting for Φ_{max} using eqn. (5.6),

$$V_s = 4{\cdot}44 f N_s \frac{V_p}{4{\cdot}44 f N_p}$$

$$= \frac{N_s}{N_p} V_p$$

thus

$$\frac{V_s}{V_p} = \frac{N_s}{N_p} \tag{5.9}$$

The ratio of secondary to primary voltage is the ratio of the secondary turns to primary turns. In practice, *i.e.* not making the assumptions above, this is a very close approximation. By adjusting the turns ratio any value of voltage may be obtained at the secondary. This is one of the principal advantages of the transformer.

Lenz's law states that the voltage induced in the secondary acts in a direction so as to oppose what is causing it. This means that when current is drawn from the secondary winding the m.m.f. set up by the secondary current will oppose the main flux causing the induction. Kirchhoff's law applied to the primary circuit states that the primary induced voltage which is also determined by the main flux is equal and opposite to the applied voltage. If the main flux were reduced by the demagnetising effect of the secondary current this would reduce the primary induced voltage and Kirchhoff's law would not hold. What in fact happens is that when secondary current is drawn a primary current flows of such a value that its m.m.f. opposes the demagnetising m.m.f. of the secondary so that the main flux remains unaffected and Kirchhoff's law is still valid. It should be remembered at this point that the m.m.f. causing the main flux is assumed negligible because of the extremely high permeability of the core. In practice, as described below, a small component of the primary current is in fact required to establish the main flux.

The demagnetising m.m.f. of the secondary winding is equal to the current-turns product, *i.e.* I_sN_s, using the symbols above. The compensating primary m.m.f. (neglecting the m.m.f. to set up the main flux) is I_pN_p.

Thus,

$$I_pN_p = I_sN_s$$

and

$$\frac{I_s}{I_p} = \frac{N_p}{N_s} \qquad (5.10)$$

which is again a close approximation when the 'ideal' assumptions are not made. Note that the secondary:primary current ratio is the primary:secondary turns ratio, *i.e.* the inverse of the voltage relationship.

It is sometimes wrongly implied that the opposing nature of the induced voltage across the secondary leads to it being in antiphase with the applied primary voltage. This is not so since the phase relationship between primary and secondary voltages of a transformer is determined by which ends of the windings are taken as reference. For example in the circuit of Fig. 5.2 the potential at point C with reference to point D is rising and falling in phase with the potential at point A with reference to B. Alternatively, the potential at point D with reference to point C is changing in antiphase with the potential at point A with reference to B. It is thus possible to obtain a zero phase shift or a 180° phase shift from primary to secondary by the appropriate connection of the windings relative to one another. The opposing nature of the induced voltages thus shows itself only in the establishment of the m.m.f. as described above and not in the phase shift, if any, between windings.

The equations of the ideal transformer may be used to show another useful characteristic, that of impedance matching. The necessity for matching between certain types of circuit was described in Chapter 3.

From eqn. (5.9)

$$V_p = \frac{N_p}{N_s} V_s \qquad (5.11)$$

From eqn. (5.10),
$$I_p = \frac{N_s}{N_p} I_s \qquad (5.12)$$

Dividing eqn. (5.11) by (5.12),

$$\frac{V_p}{I_p} = \left(\frac{N_p}{N_s}\right)^2 \frac{V_s}{I_s}$$

but
$$\frac{V_s}{I_s} = Z_s$$

(where Z_s is the secondary impedance), so that

$$\frac{V_p}{I_p} = \left(\frac{N_p}{N_s}\right)^2 Z_s \qquad (5.13)$$

In the circuit of Fig. 5.2 the ratio V_p/I_p represents the impedance into which the source generator is working. It is this effective impedance which determines the current supplied by the source, i.e. the effective impedance is the load as 'seen' by the source. So that,

$$\text{effective load} = \left(\frac{N_p}{N_s}\right)^2 Z_s \qquad (5.14)$$

The circuit of Fig. 5.2 assumed a source having zero internal impedance. In practice where the internal impedance is finite the current supplied by the source will be determined by the internal impedance and effective impedance in series (see Fig. 5.3). Determination of the effective impedance using eqn. (5.14) is known as transferring the load impedance from secondary to primary. The maximum power transfer theorem (Chapter 3) states that, for transfer of maximum power from a source to load, the source resistance and load resistance should be equal. By insertion of a transformer between a source and load the effective load resistance may be made equal to the source resistance by a suitable choice of transformer turns ratio. This method of 'matching source to load' is commonly used in power amplifier circuits.

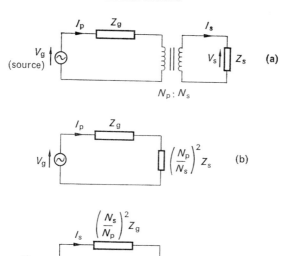

Fig. 5.3   Transformer transfer of impedances.

Note that as impedances, voltages and currents can be transferred from secondary to primary, the reverse process also applies. If all quantities in Fig. 5.3(b) are transferred to the secondary, the generator appears as $(N_s/N_p) \times V_g$, the source impedance as $(N_s/N_p)^2 Z_g$ and the load impedance would remain at Z_s.

The current I_p in the circuit of Fig. 5.3(b) is given by:

$$I_p = \frac{V_g}{Z_g + (N_p/N_s)^2 Z_s}$$

$$= \frac{N_s^2 V_g}{N_s^2 Z_g + N_p^2 Z_s} \qquad (5.15)$$

The current I_s in the circuit of Fig. 5.3(c) is given by:

$$I_s = \frac{(N_s/N_p) V_g}{(N_s/N_p)^2 Z_g + Z_s}$$

$$= \frac{N_p}{N_s} \left(\frac{N_s^2 V_g}{N_s^2 Z_g + N_p^2 Z_s} \right)$$

$$= \frac{N_p}{N_s} I_p$$

which is the result obtained in eqn. (5.12).

The validity of the circuits is thus proved and either circuit may be used in analysis.

Example 5.1
A voltage source having a sinusoidal voltage waveform of amplitude 2 V and an internal resistance of 40 Ω is connected *via* a transformer to a 4000 Ω load.
Calculate:

(*a*) the turns ratio of the transformer for maximum power transfer between source and load,

(*b*) the load current, voltage and power under these conditions.

For maximum power transfer the effective (transferred) load resistance should equal the source resistance of 40 Ω. From eqn. (5.14),

$$40 = \left(\frac{N_\text{p}}{N_\text{s}}\right)^2 4000$$

and

$$\frac{N_\text{p}}{N_\text{s}} = \left(\frac{40}{4000}\right)^{\frac{1}{2}}$$

$$= \frac{1}{10}$$

The turns ratio, primary : secondary, is thus 1 : 10.

(*b*) The secondary voltage, V_s, is given by

$$V_\text{s} = \frac{N_\text{s}}{N_\text{p}} V_\text{p}$$

$$= 10 \times 2$$

$$= 20 \text{ V (peak)}$$

The secondary current may be obtained from eqn. (5.12) after first determining the primary current using an equivalent circuit of the form shown in Fig. 5.3(*b*). Alternatively, it may be obtained directly from a circuit of the form shown in Fig. 5.4(*a*)

$$I_\text{p} = \frac{2}{80}$$

$$= 0 \cdot 025 \text{ A}$$

and from eqn. (5.12)

$$I_\text{s} = 0 \cdot 0025 \text{ A}$$

Alternatively, from Fig. 5.4(*b*)

$$I_s = \frac{20}{8000}$$

$$= 0{\cdot}0025 \text{ A}$$

as before.

The power in the load

$$= \left(\frac{0{\cdot}0025}{(2)^{\frac{1}{2}}}\right)^2 \times 4000$$

(note that the r.m.s. value must be used)

$$= 12{\cdot}5 \text{ mW}$$

This is the maximum power that it is possible to transfer from this particular source.

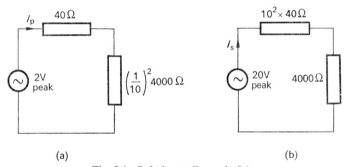

(a)   (b)

Fig. 5.4   Relating to Example 5.1.

5.4   THE PRACTICAL TRANSFORMER OFF LOAD

The ideal transformer assumes that the m.m.f. required to set up the main flux is negligible and that core losses are zero. This implies that on no load no primary current flows since a primary current is only required to set up the compensating m.m.f. for the demagnetising effect of the m.m.f. produced by the secondary (load) current.

In practice an m.m.f. is required for the main flux and the core losses are finite. On no load, then, a small primary current flows, designated I_0 in the phasor diagram of Fig. 5.5(*a*). The no load current is displaced from the applied voltage by almost $\pi/2$ rads (compare the phasor diagram of a pure inductor, Fig. 2.5(*b*)) and is almost in phase with the flux phasor. The no load current may be considered to have two components, I_c, in phase with the applied voltage, the product V_pI_c being the power loss within the magnetic core due to hysteresis and eddy currents (see below), and a component I_m in phase with and producing the main flux Φ. The primary induced voltage E_p is in phase with V_p and leads the flux causing it by $\pi/2$ rad. This diagram is somewhat simplified in that no account is taken of the primary winding impedance. (In more detail V_p is actually composed of two components, one equal and opposite to E_p and one being the voltage drop due to the finite winding impedance.)

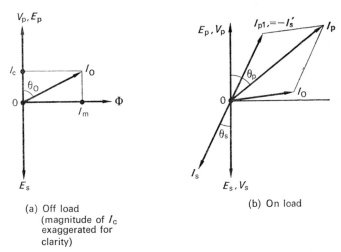

(a) Off load
(magnitude of I_c exaggerated for clarity)

(b) On load

Fig. 5.5 Simplified phasor diagram of transformer on and off load.

Note that, from the diagram,

$$I_0{}^2 = I_c{}^2 + I_m{}^2$$

and

$$I_0 = (I_c{}^2 + I_m{}^2)^{\frac{1}{2}} \tag{5.16}$$

The no load power factor

$$\cos \theta_0 = \frac{I_c}{I_0}$$

$$= \frac{I_c}{(I_c{}^2 + I_m{}^2)^{\frac{1}{2}}} \tag{5.17}$$

In the phasor diagram of Fig. 5.5(a), E_s is shown in antiphase with V_p but it can equally be in phase as was described above. For simplification the turns ratio has been taken as unity so that E_p and E_s are shown equal.

5.5 THE PRACTICAL TRANSFORMER ON LOAD

As was explained earlier the secondary current taken on load sets up an opposing m.m.f. which is then neutralised by the m.m.f. set up by the increased primary current. In the phasor diagram of Fig. 5.5(*b*) I_p is shown to consist of two components, the no load current I_0 and a component I_{p1}. I_{p1} is equal and opposite to the value of the secondary current referred to the primary, *i.e.* to $-I'_s$ where

$$I'_s = \frac{N_s}{N_p} I_s$$

This being so,

$$I'_s N_p = I_s N_s$$

and since $-I'_s = I_{p1}$

$$-I_{p1} N_p = I_s N_s$$

i.e. the m.m.f. due to I_{p1} cancels the m.m.f. due to I_s. The primary phase displacement is θ_p and the secondary phase displacement is θ_s. θ_p is determined by the relative phase of I_0 and I_{p1}. The phase of I_{p1} is determined by I'_s and I_s. The angle θ_s is determined by the reactive nature of the secondary load.

Notice that again both primary and secondary winding impedances have been neglected. In practice, as V_p has two components, so V_s, the terminal voltage at the load, is in fact the induced voltage E_s less the voltage drop due to winding impedance (phasor subtraction).

Also a unity turns ratio is assumed for convenience and again E_s is shown in antiphase with V_p. This may or may not be the case as explained earlier.

As the load current is increased, *i.e.* I_s becomes larger, the component $I_{p1} = -I'_s$ will also increase reducing the phase shifting effect of the no load current I_0 on the actual primary current I_p. For large values of load current the phase displacement and thus power factor of the primary will then become closer to that of the secondary.

Example 5.2
A single-phase transformer of primary to secondary turns ratio 8:1 takes a no-load current of 1 A at 0·2 power factor. Determine the total primary current and power factor when the transformer delivers a current of 120 A at unity power factor. Neglect the windings impedance.

The phasor diagram is shown in Fig. 5.6 for a transformer having unity power factor on the secondary side.

OD is the magnetising component I_m of the no load current I_0.
OA is the core loss component I_c of the no load current I_0.
OB is the secondary current referred to the primary I'_s.
The primary current I_p is the resultant of OB (I'_s) and I_0. The primary power factor is $\cos \theta_0$ on no load.

The solution may be obtained by accurate drawing of the phasor diagram or alternatively by resolving phasors into horizontal and vertical components. This latter method will be employed.

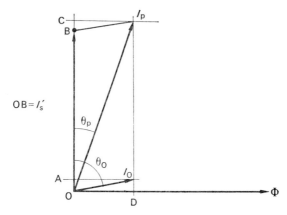

Fig. 5.6   Phasor diagram for Example 5.2.

The example gives the following information:

$$I_0 = 1 \text{ A}$$
$$\cos \theta_0 = 0\cdot2$$
$$I_s = 120$$
$$\frac{N_p}{N_s} = 8$$

It follows that:

$$I'_s = \frac{120}{8}$$
$$= 15$$
$$I_c = I_0 \cos \theta_0$$
$$= 0\cdot2$$
$$I_m = I_0 \sin \theta_0$$
$$= 0\cdot9798$$

from tables.

I_p has a vertical component

$$OC = OB + BC$$
$$= OB + OA$$
$$= I'_s + I_c$$
$$= 15\cdot2$$

and a horizontal component

$$OD = I_m$$
$$= 0\cdot9798$$

so that

$$I_p = (OD^2 + OC^2)^{\frac{1}{2}}$$
$$= (15\cdot2^2 + 0\cdot9798^2)^{\frac{1}{2}}$$
$$= 15\cdot23 \text{ A}$$

Power factor on load at the primary side

$$\cos\theta_p = \frac{OC}{I_p}$$
$$= \frac{15\cdot2}{15\cdot23}$$
$$= 0\cdot9979$$

Notice that the primary power factor closely approximates to the secondary power factor when the secondary current referred to the primary is large compared to the no load primary current.

Example 5.3
An ideal transformer having a primary winding of 2000 turns has a 25 Ω resistive load across the secondary winding. An alternating voltage of 100 V at 50 Hz input produces 75 V across the load. Calculate:

(a) the number of turns on the secondary winding,
(b) the equivalent input resistance of the transformer,
(c) the current in each winding when on load.

(a) The ratio of primary: secondary voltage

$$= \frac{100}{75}$$

Thus the ratio of primary:secondary turns

$$= \frac{100}{75}$$

Primary turns

$$= 2000$$

Therefore secondary turns

$$= \frac{75}{100} \times 2000$$

$$= 1500$$

(b) Equivalent input resistance of the transformer is the resistance 'seen' by a source connected to it, *i.e.* the load resistance referred to the primary;

$$\text{input resistance} = \left(\frac{N_p}{N_s}\right)^2 \times Z_s$$

from eqn. (5.14)

$$= \left(\frac{100}{75}\right)^2 \times 25$$

$$= 44.4 \ \Omega$$

(c)

$$\text{Secondary current} = \frac{\text{secondary voltage}}{\text{load}}$$

$$= \frac{75}{25}$$

$$= 3 \ \text{A}$$

$$\text{primary current} = \frac{N_s}{N_p} \times I_s$$

from eqn. (5.12)

$$= \frac{75}{100} \times 3$$

$$= 2.25 \ \text{A}$$

Example 5.4
A 200/100 V, 50 Hz single-phase transformer is supplied at 200 V. The primary winding has 200 turns and the core has a cross-sectional area of $50 \times 10^{-4} \ \text{m}^2$. Calculate:

(a) the number of turns on the secondary winding,
(b) the maximum flux density in the core,

(*c*) the secondary load for maximum power transfer if the supply internal resistance is 100 Ω.

(*a*) Using eqn. (5.11),

$$\frac{V_p}{V_s} = \frac{N_p}{N_s}$$

so that

$$N_s = \frac{V_s}{V_p} N_p$$

$$= \frac{100}{200} \times 200$$

$$= 100 \text{ turns}$$

(*b*) The induced voltage on the primary side

$$= 4\cdot44 f N_p \Phi_{max}$$

from eqn. (5.5).

Neglecting winding impedances this induced voltage equals the primary voltage, so that

$$V_p = 4\cdot44 f N_p \Phi_{max}$$

and

$$\Phi_{max} = \frac{V_p}{4\cdot44 \times f \times N_p}$$

$$= \frac{200}{4\cdot44 \times 50 \times 200}$$

$$= 4\cdot51 \text{ mWb}$$

The maximum flux density

$$B_{max} = \frac{\Phi_{max}}{\text{core area}}$$

$$= \frac{4\cdot51 \times 10^{-3}}{50 \times 10^{-4}} \text{ Wb/m}^2$$

$$= 0\cdot902 \text{ T}$$

(*c*) Maximum power transfer occurs when the secondary load referred to the primary side equals the source internal resistance. In this case when

$$100 = \left(\frac{N_p}{N_s}\right)^2 Z_s$$

from eqn. (5.14) so that the load

$$Z_s = \left(\frac{N_s}{N_p}\right)^2 \times 100$$

$$= \frac{100}{4}$$

$$= 25 \ \Omega$$

5.6 TRANSFORMER EFFICIENCY AND LOSSES

The efficiency of modern transformers is very high indeed, being of the order of 95–99 per cent. However, the losses can never be made entirely negligible. Transformer losses may be subdivided into two types: iron losses and copper losses.

Iron losses
These losses may be further divided into hysteresis losses and eddy current losses. The hysteresis loss is common to any magnetic material subjected to a cycling flux. When the magnetising force is reduced to zero the flux density remains finite and the magnetising force must be reversed and increased in the opposite direction to reduce the flux density to zero. The hysteresis (lagging behind) of the flux density when the magnetising force is changed is due to the shifting position of so called 'magnetic domains' within the material.

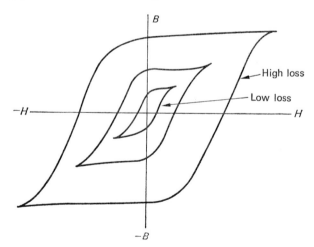

Fig. 5.7 B/H loops for various materials.

These may be visualised as small bar magnets within the magnet. If the m.m.f. varies cyclically, the continuous movement of the domains results in a loss in energy which displays itself as heat. Hysteresis loss varies with the material type as shown in Fig. 5.7 which shows several typical plots of flux density/magnetising force curves for different materials. It can be shown that the energy loss is proportional to the area contained within these curves. The curves are known as B/H loops. Hysteresis loss is proportional to the frequency of alternation.

Iron losses in general may be reduced by the choice of a low loss core material, *i.e.* having a small-area B/H loop, to reduce hysteresis losses and by laminating the core, as described earlier, to reduce and restrict the eddy current paths within the core.

The main flux within a transformer varies only slightly between no load and full load and so iron losses may normally be assumed constant. They may be measured by determination of the input power to the transformer (using a wattmeter) when the secondary circuit is *open circuited*. The copper losses, as described below, may be considered negligible under these conditions.

Copper losses
These are the normal resistance power losses caused when current flows through the windings. An alternative name for them is the I^2R loss, the name being self-explanatory. Copper losses at full load current may be measured by short circuiting the secondary and applying a low voltage to the primary of sufficient value to circulate full load secondary current. Under these conditions the iron loss is fairly small compared to the copper loss and a wattmeter connected at the input may be assumed to measure the full load copper losses only.

The efficiency of a transformer is given by:

$$\eta = \left(\frac{\text{output power}}{\text{input power}}\right) \times 100\% \qquad (5.18)$$

$$= \left(\frac{\text{input power} - \text{losses}}{\text{input power}}\right) \times 100\%$$

$$= \left(1 - \frac{\text{losses}}{\text{input power}}\right) \times 100\%$$

$$= \left(1 - \frac{\text{iron loss} + \text{primary copper loss} + \text{secondary copper loss}}{\text{input power}}\right) \times 100\% \qquad (5.19)$$

It can be shown that maximum efficiency of a transformer occurs when the (variable) copper losses are equal to the (constant) iron losses. This may be found by applying differential calculus to eqn. (5.19) when suitably expressed mathematically.

Example 5.5

A 150 kVA transformer has an iron loss of 700 W and a full load copper loss of 1800 W. Calculate:

(a) the efficiency at full load at 0·8 power factor,
(b) the maximum efficiency and the kVA output at which this occurs, assuming the same power factor.

(a)
$$\text{input power} = \text{input volt-amperes} \times \text{power factor}$$
$$= 150 \times 0\text{·}8 \text{ kW}$$
$$= 120 \text{ kW}$$
$$\text{iron losses} = 700 \text{ W}$$
$$\text{copper losses} = 1800 \text{ W}$$

Hence from eqn. (5.19),

$$\text{efficiency} = \left(1 - \frac{700 + 1800}{120\,000}\right) 100 \text{ per cent}$$

$$= (1 - 0\text{·}0208)\,100 \text{ per cent}$$
$$= 97\text{·}92 \text{ per cent}$$

(b) Maximum efficiency occurs when

$$\text{copper losses} = \text{iron losses}$$

In this case when

$$\text{copper losses} = 700 \text{ W}$$

or

$$\text{total loss} = 1400 \text{ W}$$

The efficiency under these conditions

$$= \left(1 - \frac{1400}{120\,000}\right) 100 \text{ per cent}$$

$$= 99\text{·}83 \text{ per cent}$$

Output kVA

$$= 0\text{·}9983 \times 120$$
$$= 119\text{·}796 \text{ kVA}$$

Example 5.6

A single-phase transformer is rated at 10 kVA 230/100 V. When the secondary terminals are open circuited and the primary winding is supplied at normal voltage (230 V), the current input is 2·6 A at a power factor of 0·3. When the secondary terminals are short-circuited and a reduced primary voltage causes the full load current to flow in the secondary, the primary power input is 240 W. Calculate:

(a) the efficiency at full load, unity power factor,
(b) the value of the maximum efficiency.

(a) From the open circuit test,

$$\text{iron losses} = 230 \times 2 \cdot 6 \times 0 \cdot 3$$
$$= 179 \cdot 5 \text{ W}$$

From the short circuit test,

$$\text{copper losses} = 240 \text{ W}$$
$$\text{total losses} = 419 \cdot 5 \text{ W}$$

$$\text{efficiency} = \left(1 - \frac{419 \cdot 5}{10\ 000}\right) 100 \text{ per cent}$$

from eqn. (5.19). (The input is 10 000 W since the power factor is unity)

$$= 95 \cdot 8 \text{ per cent}$$

(b) Maximum efficiency occurs when

$$\text{copper loss} = \text{iron loss}$$

i.e.

$$\text{total losses} = 2 \times \text{iron loss}$$
$$= 359 \text{ W}$$

$$\text{maximum efficiency} = \left(1 - \frac{359}{10\ 000}\right) 100 \text{ per cent}$$

$$= 96 \cdot 41 \text{ per cent}$$

PROBLEMS ON CHAPTER FIVE

(1) Discuss briefly the iron losses in a transformer and describe how they may be reduced. The primary winding of a 500 V/3000 V transformer has 100 turns and the core area is 150 cm². The supply frequency is 50 Hz. Calculate the maximum value of the flux density and the number of turns on the secondary winding.

(2) Sketch the no load phasor diagram of a transformer. What is the function of the no load current? A transformer has 100 turns on the primary winding and is connected to a 200 V, 50 Hz supply. If the no load secondary voltage is 1000 V, calculate the volts per turn of the transformer and the number of turns on the secondary winding.

(3) A single phase 600/100 V transformer takes a primary current of 1 A at a power factor of 0·3 lagging when unloaded. On load the secondary current is 12 A at 0·9 power factor lagging. Draw the phasor diagram and hence determine the primary current and the power factor.

(4) Explain why a transformer should not be connected to a dc supply.

A 400/100 V, 60 Hz transformer is connected to a 400 V supply. The primary winding has 100 turns and the core area is 100 cm². Calculate:

(a) secondary winding number of turns,
(b) the maximum core flux density.

(5) A 400 kVA, 1100/220 V single phase transformer has a 99 per cent efficiency when supplying 80 per cent full-load current at a power factor of 0·9 lagging. The copper losses and iron losses are equal under these conditions. Determine the power taken on (a) open circuit test, (b) short circuit test, and the efficiency on full load at the same power factor.

(6) A 400/200 V single-phase transformer supplies a 25 A load current at a power factor of 0·8 lagging. On no load the current and power factor are 1·5 A and 0·2 respectively. Calculate the supply current to the primary.

(7) A 500 kVA transformer has a core loss of 1 kW and a full load copper loss of 2 kW. Calculate the efficiency of the transformer at full load 0·9 power factor lagging. Determine the secondary current when the efficiency is a maximum at this power factor if the secondary voltage is 11 kV.

(8) A 10 kVA, single-phase transformer has a turns ratio of 10:1. The primary is connected to a 1000 V, 50 Hz supply. Find the open circuit secondary voltage and the approximate winding currents on full load, unity power factor.

(9) Explain briefly the significance of the readings obtained for primary power in the open and short circuit tests on a transformer. The following results were obtained on a 50 kVA transformer:

o.c. test—primary power 430 W,
s.c. test—primary power 525 W.

Calculate the transformer efficiency at full load and at half load at a power factor of 0·7 lagging.

(10) A single-phase transformer has 300 turns on the primary and 400 turns on the secondary. The transformer is connected to a 240 V, 50 Hz supply. Calculate:

(a) the no load secondary voltage,

(b) the primary current when the secondary current is 10 A at 0·8 power factor lagging, if the no load current is 0·5 A at 0·25 power factor lagging.

CHAPTER SIX

Direct Current Machines

6.1 INTRODUCTION

Electrical machines may be divided into two main categories, alternating current and direct current. These in turn may be further divided into two kinds, motors and generators. However, certain fundamental principles apply equally to all machines. This chapter firstly examines the basic principles and secondly considers in more detail how they are applied to dc machines.

All rotating machines convert energy from one form to another. The motor converts from electrical to mechanical energy, the generator converts from mechanical to electrical energy. Neither of these conversions is 100 per cent efficient since inevitably losses occur through the system.

The two basic principles on which all machine operation is based are:

(1) across any conductor situated in a changing magnetic field an e.m.f. is induced; the e.m.f. is directly proportional to the rate of change of magnetic flux;

(2) a force exists between adjacent magnetic field systems; this force depends upon the fluxes of the fields involved and may be repulsive or attractive depending on the respective polarity of the fields.

These two laws, the first of which is due to Faraday, may be used to derive the following mathematical relationships:

(i) For a conductor length l metres moving at a velocity V metres per second through a magnetic field of flux density B tesla the induced e.m.f. e is given by

$$e = BlV \text{ volts} \tag{6.1}$$

This e.m.f. is due to the conductor experiencing a changing magnetic field. The greater the value of B and V the greater is the rate of change of flux and thus the e.m.f.

(ii) For a conductor length l metres situated in a magnetic field of flux density B tesla and carrying a current i amperes the force experienced by the conductor, F, is given by

$$F = Bli \text{ newtons} \qquad (6.2)$$

This force is due to the reaction between the magnetic field, density B tesla, and the magnetic field due to the current i amperes flowing in the conductor. The greater the values of B and i the greater are the magnetic fields involved and thus the force experienced by the conductor. These equations are used in the development of e.m.f. and torque relationships derived later.

It is frequently required to determine the polarity of the induced e.m.f. or the direction of the force experienced by a current-carrying conductor in a magnetic field. Probably the easiest way to determine these is by use of Lenz's law and the fact that like fields repel and unlike fields attract. This is illustrated below.

To determine the direction of force
Examination of the fields of adjacent bar magnets as shown in Fig. 6.1(a) and (b) shows that like fields, *i.e.* those with lines of force in the *same* direction, *repel* one another and unlike fields, *i.e.* those with lines of force in the *opposite* direction, *attract* one another.

To determine polarity of induced e.m.f.
Consider a conductor moving into a magnetic field as shown in Fig. 6.1(c). Faraday's law states that an e.m.f. will be induced across

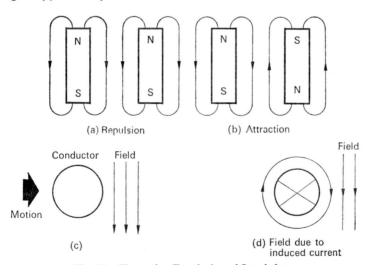

(a) Repulsion   (b) Attraction

(c)

(d) Field due to induced current

Fig. 6.1   Illustrating Faraday's and Lenz's laws.

the conductor, Lenz's law states that this induced e.m.f. will act in a direction so as to oppose what is causing it. A situation such as that illustrated in Fig. 6.1(*d*), in which current flowing in the direction shown sets up an opposing field, would result in the opposition described by Lenz. Accordingly, the induced e.m.f. acts in a direction so as to cause a current to set up an opposing field. To determine the direction all that is necessary is to decide which way the current has to flow in order that its field acts in the *same* direction as the main field causing the induction. The connection between current and field direction is covered by the familiar 'corkscrew' rule. It is suggested that this approach is easier to remember than the alternative 'right hand' or 'left hand' rules.

6.2   DC MACHINES—THE GENERAL EQUATION

In both forms of dc machines, motor and generator, the rotation of the conductors through a magnetic field results in an induced e.m.f. In the generator this e.m.f. is the driving voltage which supplies the external load. In the motor the induced e.m.f., which acts in a direction so as to oppose the applied voltage and is therefore called a 'back' e.m.f., effectively controls the machine input current according to the motor load.

The passage of current through the rotor conductors (armature) results in a voltage drop across the armature given by the product armature current times armature resistance, and the terminal voltage, the induced e.m.f. and the armature voltage drop are related to one another depending on the mode of operation of the machine. If the machine is generating, the terminal voltage is less than the induced e.m.f. by an amount equal to the armature voltage, *i.e.*

$$E = V + I_a R_a \tag{6.3}$$

where

E is the induced e.m.f. (volts)
V is the terminal voltage (volts)
I_a is the armature current (amperes)

and

R_a is the armature resistance (ohms).

If the machine is a motor the applied voltage at the armature terminals exceeds the induced e.m.f. by an amount equal to the armature voltage drop. Using the notation above, for a motor,

$$V = E + I_a R_a \tag{6.4}$$

It can be seen from eqn. (6.3) that as I_a, which in the generator case is either equal or approximately equal to the load current, falls or rises, the terminal or output voltage rises or falls (assuming constant generated e.m.f.). From eqn. (6.4), for constant applied voltage to the motor, if the induced e.m.f. rises or falls (due to the motor speeding up or slowing down) the armature current, which in this case is equal or approximately equal to the supply current, falls or rises. This interdependence of I_a, V and E is extremely important in determining regulation characteristics of a generator and torque and speed characteristics of a motor. A more detailed examination follows later.

Example 6.1
Calculate the e.m.f. generated by a separately excited dc machine providing 220 V at 20 A if the armature resistance is 0·6 Ω.

From eqn. (6.3)

$$E = V + I_a R_a \qquad (6.3)$$

thus

$$E = 220 + 20 \times 0·6$$
$$= 232 \text{ V}$$

Example 6.2
Calculate the initial rise in armature current if the back e.m.f. of a 200 V dc motor is suddenly reduced from 180 V to 150 V. The armature resistance may be taken as 0·5 Ω.

From eqn. (6.4)

$$V = E + I_a R_a \qquad (6.4)$$

thus the armature voltage drop $I_a R_a$ when the back e.m.f. is 180 V is $200 - 180$, *i.e.* 20 V, and the armature current is 20/0·5 *i.e.*, 40 A.

When the back e.m.f. falls to 150 V, the armature voltage drop initially rises by $(180 - 150)$ V, *i.e.* 30 V. The new armature voltage drop is thus 50 V.

Thus, the new armature current is 50/0·5 *i.e.* 100 A.

The initial rise in armature current is thus 60 A.

6.3 DC MACHINES—CONSTRUCTIONAL DETAILS

The simplest electrical machine consists of a single coil free to rotate between the poles of a permanent magnet. Examination of such a

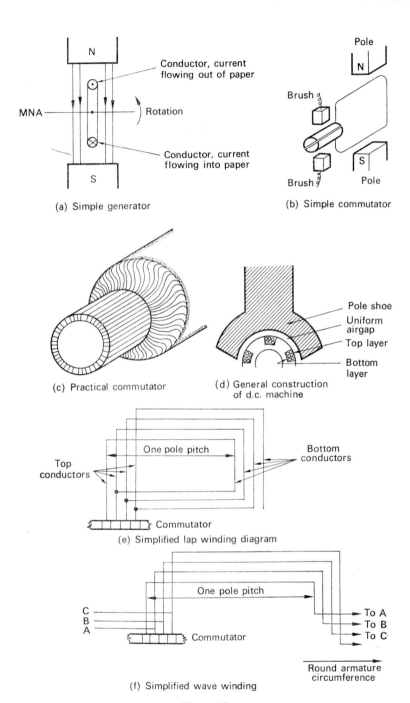

(a) Simple generator

N

Conductor, current flowing out of paper

MNA

Rotation

Conductor, current flowing into paper

S

(b) Simple commutator

Pole

N

Brush

S

Brush   Pole

(c) Practical commutator

(d) General construction of d.c. machine

Pole shoe
Uniform airgap
Top layer
Bottom layer

(e) Simplified lap winding diagram

One pole pitch

Top conductors

Bottom conductors

Commutator

(f) Simplified wave winding

One pole pitch

C
B
A

Commutator

To A
To B
To C

Round armature circumference

Figure 6.2

machine, illustrated in Fig. 6.2(*a*) shows that the induced e.m.f. fluctuates as the machine rotates. For the machine shown the e.m.f. is a maximum at the top and bottom of the diagram when the coil is *cutting* the maximum flux. (In the horizontal plane the conductors are moving *along* the flux path.) It can be seen that the e.m.f. not only changes in value but also in direction, a reversal taking place as the conductors move from the influence of the one pole to that of the other. A simple machine such as this would not be suitable for a dc machine for the following reasons:

(i) the induced e.m.f. fluctuates,
(ii) the magnetic circuit has a high reluctance due to the air gap size,
(iii) large power outputs are not obtainable,
(iv) the machine field is not controllable.

Point (i) may be taken care of by the use of a mechanical rectifier or *commutator*, as explained below. The use of coils to provide the field effectively overcomes objection (iv) (*see* Section 6.6). The remaining points are covered by the use of suitable windings.

The current from a generator based on the construction of Fig. 6.2(*a*) would alternate as the armature rotated. If the machine is used in reverse, as a motor, direct current fed into the windings would not alternate as it must in order to provide continuous unidirectional torque. Clearly what is required between the machine, within which the current alternates, and the external circuit, within which the current is unidirectional, is some form of rectifying device. A mechanical rectifier, called a *commutator* is used in dc machines.

The principle is illustrated in Fig. 6.2(*b*) which shows a simple two segment commutator. As can be seen the segment connected to a particular brush depends upon the armature position, the conductor moving under the north pole at any time (which carries current out of the paper, in the diagram shown) always being connected to the top brush and the conductor passing under the south pole always being connected to the lower brush. In this way the external circuit carries a unidirectional current whilst the actual conductor current is alternating. Thus for a generator the generated alternating current is converted to unidirectional current and for a motor the unidirectional supply current is converted to alternating current within the machine. A practical commutator has many segments as shown in Fig. 6.2(*c*). For good commutation the brushes should make contact with conductors in which there is no e.m.f. being induced. This occurs for a particular conductor when it passes through the line at which the coil current reverses direction. This is the magnetic neutral axis, denoted MNA in Fig. 6.2(*a*). The MNA actually moves when the machine is loaded due to armature reaction which is discussed in Section 6.7.

In case the reader wonders how an e.m.f. is picked up at the brushes in this case it should be noted that, in a practical machine, conductors in which there is temporarily no e.m.f. are connected to conductors in which an e.m.f. is being induced.

The two main forms of winding are lap windings and wave windings, illustrated in Fig. 6.2(*e*) and (*f*). Figure 6.2(*d*) shows the general construction of a machine utilising these windings. The conductors are laid in slots set longitudinally in a laminated magnetic core. Usually each slot contains two sets of conductors, one in the upper half of the slot, the other in the lower half. Each set constitutes one coil half, the remaining half being in a second slot situated at a distance along the drum equal to that between poles (a pole pitch). Each armature coil has a connection brought out to the commutator.

The armature as a whole rotates between poles with specially shaped end pieces (pole shoes) to give a uniform airgap as shown in the figure.

In the lap winding illustrated in Fig. 6.2(*e*) a top conductor (or set of conductors) is joined to a bottom conductor about one pole pitch away and this is then connected via a commutator segment back to the top conductor situated next to the first. In the wave winding, illustrated in Fig. 6.2(*f*) a top conductor is connected to a bottom conductor a pole pitch away, as before, but the connection is then run to a top conductor approximately a pole pitch further on, and the path runs round the armature before meeting the top conductor next to the first. In both cases the brushes are placed so that the conductors connected to them are between poles so that the current is just about to change direction. By drawing a developed diagram of the complete windings (which will not be shown here), it is possible to trace a number of separate current paths connected in parallel across the brushes. The total brush current is the sum of the currents in these parallel paths. It can be shown that for a lap winding there are as many parallel paths as there are poles in the machine, and for a wave winding there are always two parallel paths regardless of the number of poles. The number of parallel paths is an important factor in the e.m.f. equation to be considered in the next section.

6.4   THE E.M.F. EQUATION

The e.m.f. induced in a conductor cut by a changing flux is proportional to the rate of change of flux with time. For a dc machine, if

Φ is the total flux per pole (Wb)

p is the number of pole pairs

Z is the number of conductors

a is the number of parallel paths

($a = 2p$ for lap winding, $a = 2$ for a wave winding)

and N is the speed (rev/min)

the speed in revolutions per second $= (N/60)$ so that the time taken for one revolution $= (60/N)$ s.

In one revolution the conductor moves through $2p$ pole pitches; thus, the time taken to move through one pole pitch $= 60/2Np$ seconds. The flux cut during this period is Φ Wb. Therefore, the average induced e.m.f. per conductor

$$= \Phi \div \frac{60}{2Np} \text{ volts}$$

i.e.

$$= \frac{2Np\Phi}{60} \text{ volts}$$

There are Z conductors altogether and Z/a conductors in series per parallel path, thus

$$\text{total e.m.f.} = \frac{2Np\Phi}{60} \frac{Z}{a} \text{ volts} \qquad (6.5)$$

Example 6.3

A four-pole wave wound dc armature has a total of 520 conductors and is run at a speed of 950 rev/min. Calculate the induced e.m.f. when the flux per pole is 20 mWb.

Using the variable notation of eqn. (6.4) above,

$p = 2$ (number of pole pairs)

$\Phi = 20 \times 10^{-3}$ Wb

$Z = 520$

$a = 2$ (wave winding)

$N = 950$

Thus

$$\text{total induced e.m.f.} = \frac{2 \times 950 \times 2 \times 20 \times 10^{-3} \times 520}{60 \times 2}$$

$$= 329 \cdot 3 \text{ V}$$

Example 6.4

The e.m.f. generated by a certain dc machine running at 1000 rev/min with a flux per pole of 25 mWb is 500 V. Determine the new value of e.m.f. if the flux per pole is reduced to 20 mWb and the speed is increased to 1250 rev/min.

Examination of eqn. (6.5) shows that the induced e.m.f. E is given by

$$E = \text{constant} \times N\Phi$$

where the constant is determined by the machine construction. This implies that

$$E \propto N\Phi \qquad (6.6)$$

and is true, provided that the e.m.f./flux characteristic is linear, *i.e.* assuming that the magnetic circuit is not saturated.

Using the figures given and denoting the first set of conditions by subscript 1 and the second set by subscript 2

$$E_1 = 500 \text{ V}$$
$$N_1 = 1000 \text{ rev/min}$$
$$\Phi_1 = 25 \times 10^{-3} \text{ Wb}$$
$$N_2 = 1250 \text{ rev/min}$$
$$\Phi_2 = 20 \times 10^{-3}$$

so that from eqn. (6.5):

$$E_2 = \text{constant} \times 1250 \times 20 \times 10^{-3}$$

and

$$E_1 = 500 = \text{constant} \times 1000 \times 25 \times 10^{-3}$$

Thus:

$$\frac{E_2}{500} = \frac{1250}{1000} \times \frac{20 \times 10^{-3}}{25 \times 10^{-3}}$$

and

$$E_2 = 500 \text{ V}$$

The e.m.f. is thus unchanged. This is to be expected since the speed has been increased in the same ratio as the flux/pole has been reduced. Expression (6.6) is used again later when machine characteristics are discussed. Notice that the e.m.f. equation applies equally to motors and generators, the e.m.f. providing the output in a generator and being the so-called 'back' e.m.f. in a motor.

6.5   THE TORQUE EQUATION

As with the e.m.f. equation derived above, the torque equation applies equally to motors and generators. In the dc motor the torque developed by the armature provides the mechanical output whilst in the generator the developed torque opposes the driving torque, *i.e.* the energy supplied by the prime mover must have a component which is converted to sufficient rotational energy to overcome that provided by the armature torque. In a sense, this is analogous to the induced e.m.f. of a dc machine which, on the one hand, provides the electrical output of a generator but, on the other, opposes part of the electrical input to a motor. The developed torque in all dc machines is due to the reaction between the main machine field and the field set up by the armature current.

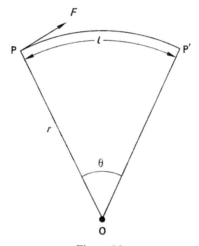

Figure 6.3

The following derivation of an expression for torque neglects machine losses and assumes that all the armature electrical power due to the induced e.m.f. and the armature current is available for conversion to mechanical power. In practice, this is not so, since in a motor part of this power is lost mechanically in friction and windage and in a generator part is lost electrically in copper losses as well as mechanical losses. (The copper losses in a motor are provided by the input power not the power due to the induced e.m.f. and the armature current; *see* Section 6.12.) The armature power as defined above, then, is given by EI_a watts where E and I_a denote e.m.f. and armature current as before.

In order to relate this power to mechanical power it is necessary to examine the meaning of the term 'torque' and its relationship to power.

Consider the diagram of Fig. 6.3 which shows a moving arm OP, length r metres, rotating about a point O. Assume that the arm has moved through an angle θ rad in t seconds, that the tangential force acting at P is F newtons and that the peripheral distance moved by point P during the time t seconds is l metres.

The torque exerted by the force, by definition, is the product of the arm length r and the force. Denoting torque by T newton-metres

$$T = Fr \text{ newton-metres} \tag{6.7}$$

The work done by the force F in moving from point P to P'

$$= Fl \tag{6.8}$$
$$= Fr\theta \text{ joules}$$

(since from geometry $l = r\theta$). The power used in the movement from P to P'

$$= \frac{Fr\theta}{t} \text{ joules per second}$$

and thus from eqn. (6.7)

$$\text{power} = \frac{T\theta}{t} \text{ watts} \tag{6.9}$$

but θ/t represents angular velocity, so that

$$\text{mechanical power} = \text{torque} \times \text{angular velocity}$$

For a machine rotating at N revolutions per minute, i.e. $N/60$ revolutions per second, the angular velocity is $2\pi N/60$ radians per second, since each revolution is equal to 2π radians, so that

$$\text{mechanical power} = T \times \frac{2\pi N}{60} \tag{6.10}$$

and neglecting losses as discussed above

$$EI_a = T\frac{2\pi N}{60}$$

Thus

$$T = \frac{60}{2\pi N} EI_a$$

From eqn. 6.5

$$E = \frac{2Np\Phi}{60} \frac{Z}{a}$$

thus,

$$T = \frac{60}{2\pi N} \frac{2Np\Phi}{60} \frac{Z}{a} I_a$$

i.e.

$$T = \frac{pZ\Phi}{\pi a} I_a \text{ newton-metres} \qquad (6.11)$$

where p is the number of pole pairs, Z is the number of conductors, Φ is the flux per pole (webers), a is the number of parallel paths, I_a is the armature current (amperes).

For a particular machine p, Z and a are fixed and

$$T \propto \Phi I_a \qquad (6.12)$$

which is true provided the e.m.f./flux relationship is linear, *i.e.* the machine magnetic circuit is not saturated. This also applied to expression (6.6).

Example 6.5
A 220 V dc motor with an armature resistance of 0·2 Ω is running at 1000 rev/min and taking an armature current of 30 A from the supply. Calculate the torque produced by the armature neglecting all losses.

From eqn. 6.4,

$$V = E + I_a R_a$$

where all symbols are as already defined, so that

$$E = V - I_a R_a$$

hence

$$\text{back e.m.f.} = 220 - 30 \times 0·2$$

$$= 214 \text{ V}$$

The armature current is 30 A so that assuming total conversion,

$$\text{mechanical power available from armature} = 214 \times 30$$

$$= 6420 \text{ W}$$

and from eqn. (6.10),

$$\text{mechanical power} = \frac{T \times 2\pi N}{60}$$

thus,

$$6420 = T \times \frac{2\pi \times 1000}{60}$$

and

$$T = \frac{60 \times 6420}{2\pi \times 1000} \text{ newton-metres}$$

$$= 61\cdot3 \text{ newton-metres}$$

Example 6.6
A four-pole lap wound dc machine with 640 conductors has an average flux per pole of 0·3 Wb. Calculate the armature torque when the armature current is 40 A.

From eqn. (6.11),

$$T = \frac{pZ\Phi}{\pi a} I_a$$

and for this problem: $p = 2$, $Z = 640$, $\Phi = 0\cdot3$, $a = 4$, $I_a = 40$. Hence

$$T = \frac{2 \times 640 \times 0\cdot3 \times 40}{\pi \times 4} \text{ newton-metres}$$

$$= 1220 \text{ newton-metres}$$

6.6   METHODS OF EXCITATION

It was stated in Section 6.3 that using a permanent magnet to provide the main field of a dc machine does not provide a controllable field. Consequently, in all but very small machines field coils are included, wound round the machine pole pieces, and the main field is adjustable by varying the field coil current. These field coils may be supplied from a separate source or connected in circuit with the armature winding, leading to the classifications *separately excited* and *self-excited* machines. To be strictly accurate, of course, only generators may be considered self-excited. If the term is applied to motors, which is unusual, it implies that the same source that supplies the motor armature also supplies the field coils.

Self-excited machines may be connected in one of three modes:

(1) so that the field coils shunt the armature,
(2) so that the field coils are in series with the armature,
(3) so that part of the field coils is in series and part in parallel with the armature.

Category 3, which is termed 'compounding' may be sub-divided into *cumulative compounding*, in which the series and shunt field coils aid one another in the setting up of the main field, and *differential compounding*, in which the series and shunt field coils oppose one

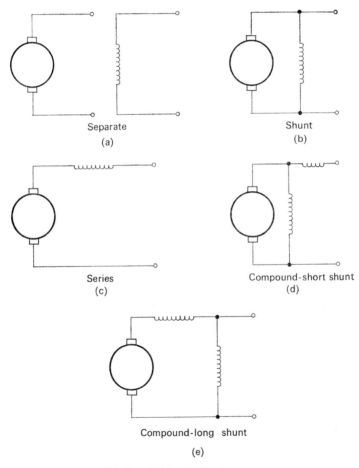

Separate
(a)

Shunt
(b)

Series
(c)

Compound-short shunt
(d)

Compound-long shunt
(e)

Fig. 6.4   Methods of excitation.

another. Compounded machines may also be connected in either a 'short shunt' or 'long shunt' mode. All forms of field connection are illustrated in Fig. 6.4. The effect of the field connection on the various machine characteristics is discussed in Section 6.9.

6.7 ARMATURE REACTION

The following phenomenon occurs equally in motors and generators, the effects on the operation depending of course on the function of the machine.

When current flows in the armature of a dc machine, a resultant magnetic flux is set up as shown in Fig. 6.5(a). There are then two fields, one due to the main field coils, shown vertically in the figure, and another due to the armature acting at right angles. The resultant field acts at an angle determined by the respective strengths of the component fields (Fig. 6.5(b)) leading to an increase in flux density

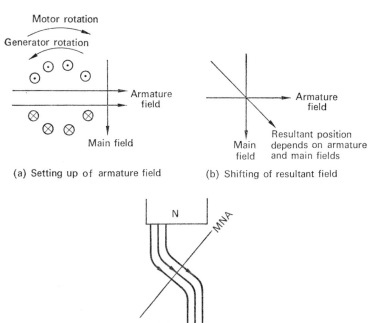

(a) Setting up of armature field   (b) Shifting of resultant field

(c) Shifting of MNA

Fig. 6.5 Armature reaction.

at one side of the pole and a decrease at the other. Since the permeability of an iron core is not linearly related to the flux density (due to the nonlinear B/H curve), the change in magnetic circuit reluctance on the weakened pole side is not balanced by the change on the strengthened pole side, and the overall effect is an increase in magnetic circuit reluctance and a consequent reduction in flux. This weakening of the main field due to armature reaction has an effect on the machine characteristics discussed in the next section.

For the current directions shown in Fig. 6.5(a), a generator would be rotating anticlockwise and a motor clockwise. It was stated in Section 6.3 that for effective commutation the brushes should be situated along the magnetic neutral axis. As can be seen in Fig. 6.5(c) the MNA moves due to armature reaction and if the brushes are left at the original position commutation worsens. One solution is to move the brush position to the new MNA (forward for generators, backward for motors), but as can be seen the MNA position is determined to some extent by the armature m.m.f. and thus the armature current, and this solution is not always effective, especially in applications involving fluctuating loads. An alternative method which also reduces armature reaction to some extent is to use interpoles placed between the main poles in a manner so that the interpole flux compensates for the armature flux. The interpole field coils are connected in series with the armature so that the interpole field is also a function of armature current, as is the armature field. Thus, the interpoles compensate for the m.m.f. produced by the armature at all load levels.

6.8 MACHINE CHARACTERISTICS

In the examination of machine characteristics it is necessary to separate machines into generators and motors and consider them separately. This is so because the characteristics of importance which are used in making a choice of machine for a particular application are clearly determined by machine function.

DC generators
The characteristics of importance in the case of the generator are

(a) e.m.f. versus speed,
(b) e.m.f. versus field current,
(c) e.m.f. versus armature (load) current.

These will now be considered in turn for the various forms of generator.

Separately excited generators

The e.m.f. versus speed characteristic of a separately excited generator is shown in Fig. 6.6(a). It follows from expr. (6.6)

$$E \propto N\Phi \qquad (6.6)$$

that for fixed field current and thus a fixed value of Φ, $E \propto N$ which yields the linear relationship shown. Curve 2 is for a higher (fixed) value of I_f than curve 1. If the speed is held constant

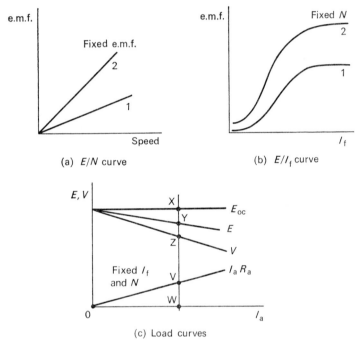

(a) E/N curve

(b) E/I_f curve

(c) Load curves

Fig. 6.6 Characteristics of separately excited dc generators.

then, from above, $E \propto \Phi$. The relationship between Φ and I_f is not linear but follows the B/H curve for the magnetic circuit and so the plot of E versus I_f also follows the B/H curve. Curve 2 is for a higher (fixed) speed than curve 1. The initial value of the e.m.f. at zero field current is due to the remanent magnetism of the magnetic circuit and is higher as the speed is increased.

On load, *i.e.* for increasing values of armature current, the main flux is weakened owing to armature reaction and the graph of E

versus I_a for fixed I_f and N shows a droop as I_a is increased. The terminal voltage V is related to the e.m.f. E by eqn. (6.3)

$$V = E - I_a R_a \qquad (6.3)$$

so that for the curve shown in Fig. 6.6(c) at any value of I_a, OW say,

$$WV = YZ = I_a R_a$$

and

$$WZ = WY - YZ$$

The drop XY is due to armature reaction.

Curves (a) and (b) of Fig. 6.6 are called *open circuit characteristics* and curve (c) the *load characteristic* for the machine.

Shunt excited generators

A shunt excited generator provides its own field current, the initial excitation being provided by the residual magnetism of the magnetic circuit. The small e.m.f. generated due to this remanent flux sets up a small field current which in turn increases the main field and thus the induced e.m.f. The process continues until saturation is reached when the e.m.f. remains substantially constant. If for some reason there is no residual magnetism, *e.g.* with a new machine, or the residual flux is in the wrong direction the machine will fail to excite and thus will not generate. The e.m.f. versus speed curve is shown in Fig. 6.7(a). It can be seen that the e.m.f. rises only slowly until a certain speed, indicated by N_c, is reached at which point generation begins in earnest. N_c is called the critical speed of the machine. To examine why the shape of the curve is as shown, it is necessary to take a closer look at the e.m.f./field current curve illustrated in Fig. 6.7(b).

This figure shows two graphs, the e.m.f./field current curve which has the familiar shape of Fig. 6.6(b), and for the same reasons as outlined above, and a plot of the voltage drop across the field windings due to field current and field coil resistance against field current. As can be seen this graph is linear since the voltage is directly proportional to I_f and the field resistance R_f is constant. Until point W is reached, at which the e.m.f. and the drop $I_f R_f$ are equal, the e.m.f. exceeds the field voltage by an amount which varies in magnitude, being a maximum at point X. At this point the total e.m.f. is XZ, the $I_f R_f$ drop is YZ and the excess voltage is XY.

Since the field coils are of necessity highly inductive, as long as the field current is changing there will be a back e.m.f. induced in them which is proportional to the rate of change of flux in the magnetic circuit. The difference in voltage between the generated e.m.f. E and the field (resistive) voltage drop $I_f R_f$ is equal and opposite to this back e.m.f. At point W when the e.m.f. is equal to the field voltage drop, the flux has stopped changing and there is no back e.m.f. induced in the field coils. If the resistance of the field coils is increased, a greater proportion of the generated e.m.f. is required to provide any particular value of field current, and less e.m.f. is therefore

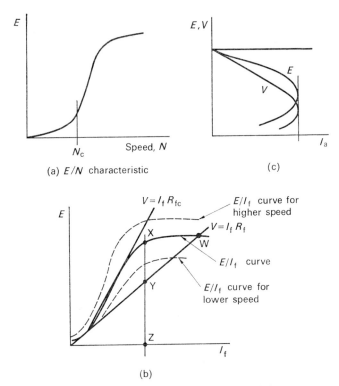

(a) E/N characteristic

(c)

(b)

Fig. 6.7   Shunt excited generator characteristics.

available to oppose the back e.m.f. of self inductance. Thus the point at which the field current stops increasing is brought further down the E/I_f characteristic. If the field resistance is too high, *i.e.* the $I_f R_f/I_f$ graph lies to the left of the E/I_f curve, the field current cannot increase since all the e.m.f. available is required for the $I_f R_f$ drop, and there is no component available to counteract the back e.m.f. Since I_f cannot increase, the machine fails to excite. At a certain value of field resistance R_{fc}, giving the graph $V = I_f R_{fc}$ in Fig. 6.7(b), conditions are unstable, the machine being at the point of being able to excite. A slight reduction of R_f at this point would lead to immediate excitation. The value R_{f_c} is called the *critical resistance* of the machine. If the field resistance is higher than this value the machine will not excite. The curve of Fig. 6.7(b) plotting E against I_f is for a constant speed. Broken lines are shown for a higher value and for a lower value of speed. Examination of these curves shows that the graph of $I_f R_f$ which is tangential to each curve of E/I_f has a different slope, being lower for the lower speed and higher for the higher speed. Since the tangential graph of $I_f R_f$ gives the critical resistance, this implies that there is a different value of critical resistance for each value of speed. The critical resistance (given by the *slope* of the tangential $I_f R_f$ graph) is *lower* for a *lower* speed and *higher* for a *higher* speed. As any value of speed has associated with it a critical resistance so any value of resistance has associated with it a speed at which this resistance is critical. This speed is called the *critical speed*. At speeds below the critical speed, the resistance is higher than the critical resistance for these lower speeds and the machine will not excite. At speeds higher than the critical speed, the resistance is lower than the critical resistance for these higher speeds and the machine excites and thus generates. At the critical speed for any particular field resistance the machine is unstable. The e.m.f. versus speed curve is shown in Fig. 6.7(a), the critical speed for this value of field resistance being shown as N_c.

The load characteristic of a shunt generator is shown in Fig. 6.7(c). An unusual feature of this curve is the manner in which the e.m.f. curve and thus the terminal voltage curve turn back once a certain level of armature current is reached. This again is due to the fact that the machine provides its own excitation. At a certain value of load current, mainly determined by the machine magnetic circuit, the terminal voltage V has been so far reduced, owing to the combined effects of armature reaction and the armature voltage drop due to resistance, that the machine is no longer capable of sustaining the excitation required. Any attempt to draw more than this critical value of armature current brings the operating point to the lower part of the curve.

Series excited generators
In the series excited generator, the load (armature) current is also the excitation current, and graphs of E and V against I_a are also those of E and V plotted against I_f.

For a constant excitation current the e.m.f. is proportional to speed and the E versus N characteristic is approximately linear, as in Fig. 6.6(a). Figure 6.8 shows the E v/s I_a (I_f) characteristics. The shape follows the familiar E v/s I_f relationship (approximately that of the

B/H curve), but it is important to note that, in this case, I_f is in fact I_a. Thus, for any operating point X, the line OX in Fig. 6.8 has a slope equal to the total load and field circuit resistance. As with the shunt generator, if this line OX lies to the left of the curve, the machine will not excite. There is thus a critical load resistance for a series machine which as before is dependent on speed. The series machine is not widely used, owing to the shape of the load characteristic and the widely fluctuating output voltage at various levels of load current.

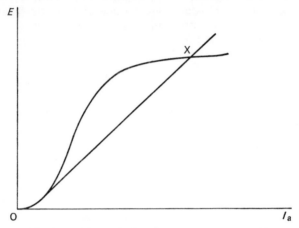

Fig. 6.8 Series excited generator E/I_f characteristic.

Compound generators
Figure 6.9 shows the terminal voltage load current characteristic of various forms of cumulative compounded generator. The overall effect depends as one might imagine on the relative effect of the series coils compared to the shunt coils. As the m.m.f. of the series coils is increased the droop of the shunt curve (Fig. 6.7(c)) is counter-acted by the addition of cumulatively connected series coils (Fig. 6.9), curves 1 to 4 showing the effect of progressively increasing the number of series turns.

DC motors
The characteristics of importance in the case of the motor are:

    (a) speed versus armature current,
    (b) torque versus armature current,
    (c) speed versus torque.

These will be considered in turn for the various types of motor.

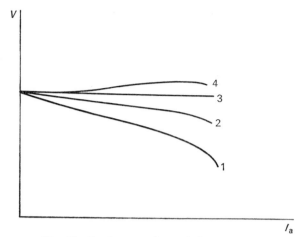

Fig. 6.9   Load curves of cumulative generator.

Shunt motor

The characteristics for a shunt motor are shown in Fig. 6.10(a). Expression (6.12) states that

$$T \propto \Phi I_a \qquad (6.12)$$

where T, Φ and I_a represent torque, flux and armature current respectively. For a shunt motor the field coils are supplied from the applied voltage, and neglecting the effects of armature reaction the flux is substantially constant, so that

$$T \propto I_a$$

which yields an approximate linear relationship, as shown in Fig. 6.10(a).

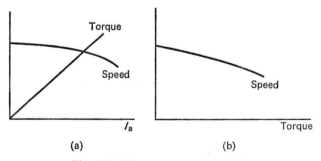

Fig. 6.10   Shunt motor characteristics.

From expr. (6.6), the induced e.m.f. in a motor

$$E \propto N\Phi \tag{6.6}$$

where N represents speed. Thus, the speed

$$N \propto \frac{E}{\Phi}$$

and from eqn. (6.4),

$$E = V - I_a R_a \tag{6.4}$$

so that

$$N \propto \frac{V - I_a R_a}{\Phi} \tag{6.13}$$

The numerator of the right-hand side of expr. (6.13) will reduce as I_a rises; the denominator will also be reduced slightly owing to armature reaction. The armature voltage drop has a somewhat greater effect than the flux reduction so that the speed characteristic droops with increasing armature current as shown in Fig. 6.10(b). However, the shunt motor may be considered a substantially constant speed machine.

The speed/torque characteristic, which may be deduced from Fig. 6.10(a) is shown in Fig. 6.10(b).

Series motor
Figure 6.11 shows the characteristics of the series connected motor. In this circuit the field coils and armature coils are in series, the armature current producing the main field.

From expr. (6.12), the torque

$$T \propto \Phi I_a \tag{6.12}$$

and since the flux Φ is produced by I_a according to a relationship approximating to the B/H relationship, before saturation Φ is

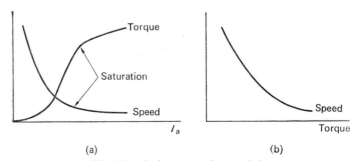

(a)   (b)

Fig. 6.11   Series motor characteristics.

approximately proportional to I_a and T is approximately proportional to $I_a{}^2$; after saturation Φ is constant and $T \propto I_a$. The torque/armature current curve is shown in Fig. 6.11(a).

From expr. (6.13), the speed

$$N \propto \frac{V - I_a R_a}{\Phi} \tag{6.13}$$

As I_a rises, the flux Φ also rises and the speed falls. If the effect of $I_a R_a$ was neglected the curve would be a rectangular hyperbola, *i.e.* that produced by a relationship

$$N \propto \frac{1}{\Phi}$$

The armature voltage drop when taken into account lowers the curve slightly in the vertical plane. The shape is also modified by the fact that flux is not directly proportional to the armature current because of saturation. Note that the speed of a series motor is far from constant with load and rises to a high value on no load (low values of I_a). The speed armature current curve is shown in Fig. 6.11(a).

The speed/torque curve which may be deduced from Fig. 6.11(a) is shown in Fig. 6.11(b).

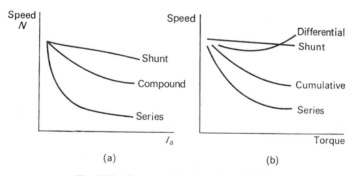

Fig. 6.12 Compound motor characteristics.

Compound motor
As with the compound generator, the characteristics of a compound motor are a compromise between those of the series and shunt motors. The exact shape depends on the relative effect of the series winding compared with the shunt winding (*see* Fig. 6.12).

6.9   APPLICATIONS

Generators
Examination of Figs. 6.6–6.9 indicates that the series generator is unsuitable for use as a supply generator on its own, although it may be used in conjunction with a substantially constant voltage machine (shunt or compound) as compensation for increased voltage drop due to increasing load current. For a small system a compound generator alone, which embodies the principle of shunt plus series in one machine, can be used. An over-compounded generator is especially useful from this point of view.

The shunt or separately excited generator is the most commonly used machine for local supply systems (within vehicles, etc.), the disadvantage of the latter usually being overcome by mounting a small shunt generator on the same shaft as the large separately excited machine to provide the field current.

Motors
As was shown earlier, the shunt motor runs at fairly constant speed over a wide load range. As such it is most useful for a number of low load applications. It would not be suitable for fluctuating loads, such as, for example, electrically powered vehicles. For this use the series machine with the falling speed characteristic is ideal since as the speed and hence the back e.m.f. falls, the armature current and thus the torque rises (eqns. (6.4) and (6.11)). A series motor runs at extremely high speeds on no load and is therefore never coupled mechanically to a load in any manner which could result in load disconnection (belt drives, etc.). The compound motor which has a stable no load speed but retains the series speed curve is a useful alternative to the series machine in certain applications, *e.g.* steel mill drive units.

6.10   MOTOR STARTERS

Equation (6.4) shows that the difference between the applied voltage and the armature voltage drop is the back e.m.f., which depends, of course, upon the motor speed. The back e.m.f., in fact, acts as a control on the armature current in that, as the one falls, the other rises. The bulk of the applied voltage is used to overcome the back e.m.f. On first closing the supply circuit there is no back e.m.f. since the machine is not turning, and without any additional resistance the armature current would rise to a very high value. For example, a 200 V 0·5 Ω machine would have an initial current of 400 A.

Obviously, such high currents are undesirable and a motor starter is employed in which a series resistor is progressively reduced as the motor speeds up, the increasing back e.m.f. taking over control of the armature current. A dc motor starter should always be operated smoothly, neither too quickly or too slowly. A typical starter circuit is shown in Fig. 6.13.

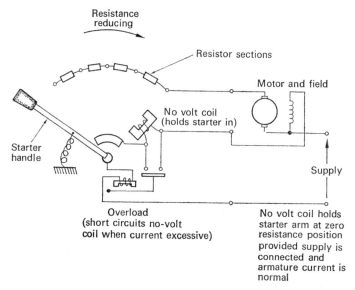

Fig. 6.13   Motor starter.

6.11   CONTROL OF MOTOR SPEED

It can be seen from expr. (6.13) that motor speed depends upon applied voltage and the machine flux. This offers two methods of speed control. For the shunt or compound machine the speed may be adjusted upwards from that at normal excitation by the addition of a series resistor called a *field regulator* as shown in Fig. 6.14(*a*). Increasing the field regulator resistance reduces the field current and thus the machine flux. Hence the speed rises. The speed cannot of course be adjusted to a value below that at normal excitation using this method. For speeds from zero to the normal speed, the method illustrated in Fig. 6.14(*b*) is employed, in which a series resistor, called a controller, is included with the motor across the supply voltage. Increasing this controller resistance reduces the applied voltage to the motor. With this method speeds from that at normal excitation down to zero may be obtained.

Two methods of controlling a series machine are shown in Figs. 6.14(*c*) and (*d*). Figure 6.14(*c*) shows the use of a shunt resistor across the field coils, called a diverter. Minimum speed is obtained with the diverter circuit open, so that the method is used to obtain speeds above that with normal excitation. Figure 6.14(*d*) shows the use of a tapped field coil. Adjustment of the total number of turns varies the m.m.f. and thus the flux for a given load current.

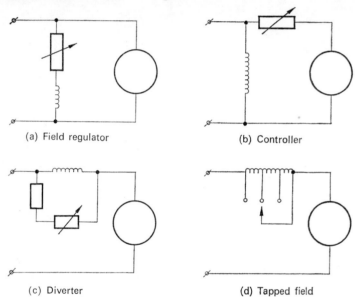

(a) Field regulator

(b) Controller

(c) Diverter

(d) Tapped field

Fig. 6.14   Motor speed control.

6.12   MACHINE LOSSES

All forms of dc machine may be regarded as energy convertors. Certain losses inherent in the machine are common to both motors and generators. These are:

(*a*) magnetic circuit iron loss due to hysteresis and eddy currents,

(*b*) copper losses in the armature and field windings due to the coil resistance,

(*c*) friction and windage losses due to the bearings and the setting up of air currents within the machine,

(*d*) brush contact resistance losses at the commutator.

For a generator, multiplication of eqn. (6.3) by the armature current, I_a, yields

$$EI_a = VI_a + I_a{}^2 R_a \tag{6.14}$$

EI_a represents the available power from the armature which yields VI_a, the output (electrical) power plus the friction and windage loss (and field losses if the machine is self-excited) and $I_a{}^2 R_a$, the armature copper loss. For a motor, multiplication of eqn. (6.4) by I_a yields

$$VI_a = EI_a + I_a{}^2 R_a \tag{6.15}$$

VI_a is the input (electrical) power to the armature provided by the supply, which yields EI_a, the output (mechanical) power plus friction and windage losses, and $I_a{}^2 R_a$, the armature copper loss. The field losses may also be included in the VI_a term if the machine is series connected or may constitute a separate term supplied in addition to VI_a if shunt excitation is employed.

PROBLEMS ON CHAPTER SIX

(1) The magnetisation curve of a dc generator is given by

Voltage	8	108	200	260	290	300	305
Field current	0	0·4	0·8	1·2	1·6	2·0	2·4

Plot the curve and deduce the voltage generated if the machine is operated as a shunt generator with a 200 Ω total field resistance.

(2) A dc shunt machine has armature and field resistance of 0·02 Ω and 100 Ω respectively. As a generator running at 400 rev/min it delivers 100 kW at a constant voltage of 350 V. Calculate the speed of the machine when supplied with 350 V and 100 kW input power.

(3) A short shunt compound generator has armature, shunt-field and series-field resistances of 1 Ω, 40 Ω and 0·5 Ω respectively, and supplies a load of 5 kW at 200 V. Calculate the armature generated voltage.

(4) A shunt machine driven at 1000 rev/min produces a direct voltage supply of 200 V at 50 A. If the machine runs as a motor with the same terminal voltage and an armature current of 50 A calculate the speed. Armature circuit resistance is 0·4 Ω.

(5) Calculate the torque developed by a 400 V dc motor with an armature resistance of 0·5 Ω, running at 700 rev/min and taking an armature current of 40 A.

(6) A dc shunt motor running at 1000 rev/min takes 25 A armature current from a 400 V supply. When 200 V is supplied the armature

current falls to 20 A and the flux is reduced to 0·8 of the original value. The armature resistance is 0·5 Ω. Calculate the speed when 200 V is applied.

(7) A shunt wound generator supplies a current of 60 A at 250 V terminal voltage. The armature resistance is 0·05 Ω and the total shunt resistance is 150 Ω. The flux per pole is 60 mWb and there are 500 conductors in the lap wound armature. Calculate the machine speed.

(8) A shunt generator supplies 1 kW to a load at 200 V terminal voltage. The armature resistance is 0·5 Ω and the field resistance 150 Ω. Calculate the generated voltage and plot the voltage regulation curve for loads from zero to 1 kW.

(9) A certain lap wound dc motor has 424 conductors and a useful flux per pole of 0·036 Wb. Determine the machine torque for a load current of 100 A.

(10) The torque from a 6-pole dc motor having a useful flux per pole of 0·04 Wb is 1865 Nm when a current of 200 A is taken from the supply. Calculate the number of conductors in the machine assuming two parallel paths.

CHAPTER SEVEN

Amplifiers and Amplifying Devices

7.1 INTRODUCTION: AMPLIFIER TYPES

An amplifier may be broadly defined as a circuit system which enlarges an electrical signal. The system may be entirely electrical or electromagnetic, in which case it may be considered to be a static system, or may be made up of a combination of electrical, or electromagnetic, and mechanical devices, in which case the system may be considered to be a dynamic system. Amplifiers, in general, may be categorised either by function or by the nature of their operating parts. Functional categories are:

(*a*) voltage, current or power,

and also

(*b*) direct coupled, audio frequency, radio frequency, and wide band;

so that any one of (*a*) may be combined with any one of (*b*) to give, for example, an audio-frequency power amplifier, which is used to drive a loudspeaker in a radio or television receiver, or a direct coupled voltage amplifier, which may be used to amplify small, slowly changing signals from a transducer in, say, an industrial control system.

Alternatively, the categories according to the nature of the amplifying device, are

(*c*) electronic, magnetic and rotating (mechanical), and a complete description of any amplifier would then require one of the categories from each of (*a*), (*b*) and (*c*) respectively. It should be noted, however, that not all combinations are possible.

Categories listed under (*a*) describe the type of signal to be amplified. It will be appreciated that, in any electrical circuit, voltages and currents are present, and thus there must also be a power level. Clearly, then if voltage or current is changed in magnitude, this will affect the levels of the other quantities. Nevertheless, the category 'voltage, current or power' refers to the *specific* quantity which it is primarily desired to amplify. Certain amplifiers amplify the one (voltage or current) but not the other and, again, in other types both are amplified, yielding a substantial power gain. Whether a voltage, current or power amplifier is required is determined by the input (driving) circuit and the output (driven) circuit. Some examples are the input amplifier of a radio receiver, which is a voltage amplifier, a temperature compensating feedback amplifier in a transistorised control system, which is a current amplifier and, finally, a servo-control amplifier used in automatic control systems, which is a power amplifier.

Categories listed under (*b*) may be further described as follows.

Direct coupled amplifiers

These amplifiers are characterised by their lack of frequency sensitive components. They handle so-called dc signals, *i.e.* a relatively slow change in the dc level of a voltage or current. A sophisticated development of the dc amplifier is the operational amplifier, used particularly in analogue computing or simulating systems.

Audio frequency amplifiers

These amplifiers are designed to amplify electrical signals at frequencies corresponding to the audible range, that is between limits of 50 Hz to 20 kHz. The reason for the words 'corresponding to' is that electrical signals, as such, are of course inaudible and a transducer (loudspeaker, headphones, etc.) is necessary for conversion into mechanical vibrations.

Radio frequency amplifiers

These amplifiers are designed to amplify narrow bands of frequencies within the radio frequency range. In general they tend to be sensitive to a particular radio frequency (usually the carrier) and are able to handle a small range either side of the centre frequency. To have this characteristic, radio frequency amplifiers use frequency selective *LC* circuits which are adjusted (tuned) to give maximum performance at the desired frequency. They are used, as the name implies, in radio and television receivers and transmitters.

Wide band amplifiers

Many electrical signals especially in control systems and television transmission systems are of the pulse variety, *i.e.* the level of voltage (or current) rises sharply to a new value and after a certain length of time falls again equally sharply to a different value. Analysis of such waveforms shows that the frequency content, *i.e.* the frequencies of the many components making up the waveform covers a very wide range lying between virtual dc and frequencies up to several megahertz.

Pulse amplifiers must therefore be able to handle with equal efficiency (as regards gain and distortion of waveshape) frequencies within this range. Such amplifiers are known as wide band amplifiers because of this wide frequency range; alternatively, in television systems they are referred to as *video* amplifiers. Special compensation techniques are necessary in such amplifiers to avoid fall off in gain of the frequencies at the extremities of the operational range.

Categories listed under (*c*) may be further discussed as follows.

Electronic amplifiers
By far the bulk of amplifier circuits employ electronic techniques using thermionic and other valves or semiconductor devices such as diodes, transistors etc. or a combination of both. Electronic amplifiers have the advantage of small physical size and the absence of moving parts and quite complex operations may be carried out using a minimum of working space. They are used extensively in the amplification of voltage, current and power signals, the range of power handling ability being somewhat restricted compared to the other two categories.

Magnetic and rotational amplifiers
Magnetic and rotational amplifiers are particularly useful for power amplification, the former having the advantage of being static (non-moving) devices, the latter having the advantage of being able to provide torque and thus mechanical energy at the output.

All categories of amplifier including the nature and characteristics of operation will be discussed more fully in the remainder of the chapter.

7.2 AMPLIFIER CHARACTERISTICS

There are several characteristics of any amplifier by which its performance may be judged and hence a comparison made for selection to carry out a particular task. Not all the characteristics are equally important or even relevant, the importance or relevance being determined largely by the type of amplifier and its function.

The main characteristics, which are discussed below, are:

(a) input and output impedances,
(b) voltage, current or power gain,
(c) transfer characteristic,
(d) response time.

Input and output impedances

The input impedance of any amplifier is the ratio between input voltage and current and is often referred to as the impedance 'seen' at the input. The output impedance, similarly, is the ratio between output voltage and output current and, again, is often termed the impedance 'looking back' into the output terminals. A simple equivalent circuit of an amplifier using the concept of input and output impedances is shown in Fig. 7.1(a). It can be seen that the

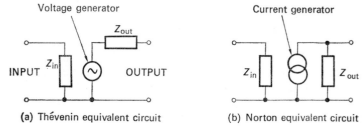

(a) Thévenin equivalent circuit   (b) Norton equivalent circuit

Fig. 7.1   Input and output impedances.

output section utilises Thévenins theorem (*see* Chapter 3) as applied to ac circuits. Alternatively, Norton's theorem may be applied to yield the circuit of Fig. 7.1(b). In both circuits Z_{in} is the 'input' impedance and Z_{out} the 'output' impedance.

The importance of input and output impedances lies in the fact that for optimum performance amplifier stages should be suitably matched, the nature of the matching depending upon the nature of the signal to be amplified. For transfer of voltage signals, for example, the input impedance of the amplifier should be as high as possible so that maximum voltage input should be developed. For transfer of signals where the current is of prime importance the input impedance should be low so that maximum signal input current might flow. The words 'low' and 'high' are, of course, relative and in this case mean with reference to the output impedance of the device which is feeding the amplifier. For a power amplifier, the maximum power transfer theorem (Chapter 3) indicates that the output impedance of the driver circuit should equal the input impedance of the amplifier.

Voltage, current or power gain

The gain of an amplifier is the ratio of the value of the output signal to that of the input signal. The importance of this characteristic is self explanatory. In complete systems it must always be borne in mind that the gain of an individual amplifier isolated from the system may be considerably modified when the amplifier is included in the system. This is one factor determined largely by the efficiency of matching as discussed above.

Transfer characteristic

The transfer characteristic of an amplifier is a graph plotting the value of the output signal against the corresponding value of input signal. The nature of the signals may be voltage, current or power as before. In general the slope of the transfer characteristic determines

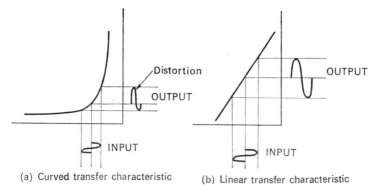

(a) Curved transfer characteristic   (b) Linear transfer characteristic

Fig. 7.2   Transfer curves.

the gain of the amplifier and the shape determines the degree of distortion of the input waveform as it is transferred through the amplifier. This latter point is clarified in Fig. 7.2 which shows the effect on a sinusoidal input passing through an amplifier with (*a*) a curved transfer characteristic and (*b*) a linear transfer characteristic.

Response time

The response time of an amplifier is a measure of the time required for it to respond to an input signal. It is an important characteristic of amplifiers to be incorporated in control systems where it is essential to know the delay which will occur between receipt of a control signal and the resultant adjustment of the device being controlled.

All the above characteristics are dependent upon frequency and it is often necessary to include special compensation circuits to counter-act the change in the characteristic at the extreme values of the frequency range covered by the amplifier.

The characteristics are also dependent to some extent on ambient temperature, particularly in the case of solid-state amplifiers, and occasionally on other physical operating conditions (air pressure, humidity, etc.).

7.3   ELECTRONIC AMPLIFYING DEVICES

Amplifying devices used in electronic amplifiers may be divided into two categories: (a) thermionic valves (or tubes) and (b) solid-state devices or transistors. In both types the same principle is used, that of controlling an electric current by means of a small signal which changes the ease with which the main current flows through the device. In the valve the current passes through an evacuated or gas-filled glass envelope, in the transistor the current passes through a piece of specially prepared material known as a semiconductor. Valves and transistors are described more fully below.

Electronic valves
Valves may be categorised as thermionic or cold cathode and as vacuum or gas-filled. Valves used for amplifying are generally thermionic vacuum types and so other varieties will not be discussed. In the amplifying valve, the glass envelope is evacuated and the current consists of a stream of electrons released from a heated cathode, which are attracted to a positive electrode called the anode. The heat necessary is provided by a filament through which current is passed from a separate source of energy. In a *directly heated* valve the heater filament also acts as a cathode and emits electrons; in the *indirectly heated* valve the cathode is a separate electrode wrapped around the heater. In both cases, the cathode is specially coated with a material which releases electrons on the receipt of suitable energy. The electrons pass only from cathode to anode and so all valves are one-way or rectifier devices.

The simplest valve consists of heater, cathode and anode and is called a *diode*. A typical construction and current–voltage characteristic is shown in Fig. 7.3, which also shows the standard symbol used in circuit diagrams. A diode is used purely for rectification purposes and not as an amplifier.

The simplest amplifier valve is a *triode*, which has a third electrode known as a grid. The grid consists of a fine wire mesh wrapped close to the cathode and the electrons must pass through it in order to get to the anode. If the grid is held at a negative potential with respect to the cathode, the electric field set up controls the electron flow between cathode and anode. If the grid is held sufficiently negative, the electron flow will stop altogether since the electrons have insufficient energy to penetrate the field. Since the grid is situated physically close to the cathode, only small voltages are necessary at the grid to yield substantial control over the anode current. Typical

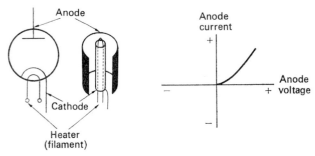

(a) Symbol   (b) Construction   (c) Anode characteristic

Fig. 7.3   The thermionic diode.

triode characteristics and construction are shown in Fig. 7.4. As can be seen, for a particular value of anode voltage, making the grid more negative reduces the anode current, making the grid more positive increases the anode current.

Further developments of the triode include the tetrode and the pentode which have an additional one or two grids, respectively, situated between the first or control grid and the anode. The development of these valves is discussed in more detail in books devoted

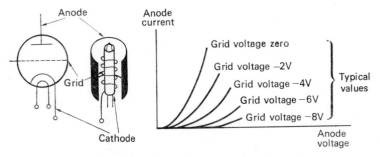

(a) Symbol   (b) Construction   (c) Anode characteristics

Fig. 7.4   The thermionic triode.

exclusively to electronics. For the purposes of this text, it is sufficient to say that the tetrode extra grid, known as the *screen* grid, reduces the capacitance between anode and control grid, thereby reducing signal feedback; the pentode extra grid, called the *suppressor*, reduces the effect of the secondary emission of electrons from the anode due to bombardment by the main stream of primary electrons. Typical characteristics and construction of a pentode valve are shown in Fig. 7.5. Triodes, tetrodes and pentodes may be used in amplifier circuits, the choice of valve type being determined by the amplifier function. Electronic amplifier circuits are described in more detail in the next section.

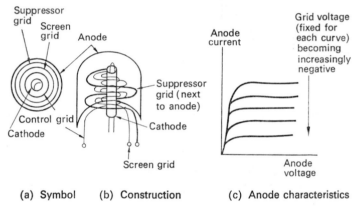

(a) Symbol   (b) Construction   (c) Anode characteristics

Fig. 7.5   The thermionic pentode.

Transistors

Transistors are made from material having conductive properties which lie between those of a good conductor such as copper and an insulator such as rubber or porcelain. These materials, two of the most widely used of which are germanium and silicon, are thus known as *semiconductors*. Both germanium and silicon are of crystalline form with tightly bound electrons and as such permit only extremely small currents to flow under the influence of an applied p.d. If suitable impurities are added in stringently controlled conditions, the pure or intrinsic semiconductor may be converted to an impure or extrinsic semiconductor of one of two types, *n*-type or *p*-type. In *n*-type silicon or germanium the impurity is such that electrons, which are more loosely bound than in the pure material, are made available for conduction. In *p*-type material, the impurity is such that electron vacancies or 'holes' appear in the material, these holes readily accepting electrons if any are available. A crystal of either p-type or n-type material has a higher conductivity than the intrinsic material due to the addition of these impurities and it should be noted that a single crystal of either type conducts equally well in both directions.

However, if p-type and n-type materials exist in close proximity, to form a pn junction, conditions are set up which allow familiar diode operation. Briefly, what happens is this: electrons moving at random, due to thermal agitation, cross from the n-type material, where they are in the majority, to the p-type material, where they are in the minority. Since initially both materials are electrically neutral, the transfer of electrons from n to p types causes the p-type to become progressively negative and a p.d. is set up across the junction. The junction p.d. tends to repel further random flow and a potential 'barrier' exists across the junction. If an external p.d. is applied to the junction so that the junction p.d. is opposed, electrons continue to flow from n to p and current flows. If, on the other hand, the external p.d. is applied to assist the barrier p.d., electrons do not flow from n to p types and a large current does not flow. Current flow does not actually cease since there are some electrons in the p-type caused by thermal generation, which move to the positively connected n-type. Since these electrons are few and are in the minority in the p-type, the small current flowing when the diode is reverse connected is called minority carrier current (*see* Fig. 7.6).

The properties of the pn junction are utilised in the following manner. In the *bipolar transistor* three layers of doped material are arranged as in Fig. 7.7(a). The emitter-base junction is forward biased and the base-collector junction is reverse biased. For the transistor shown, which is an *npn* junction transistor, the emitter majority carriers, electrons, are swept into the base region under the influence of the first junction external p.d. The base region is deliberately made very narrow so that the bulk of the carriers sweep across the base and are attracted to the collector, which for the *npn* transistor shown, is held positive. Some of the electrons leave the base region via the base lead. The situation is now that three currents exist: the emitter current, the base current and the collector current. Altering the base current by adjustment of the base-emitter bias results in a larger alteration of the collector current and so there exists the necessary condition for amplification, the ability to control a through current by the application of a small input signal.

The other form of bipolar transistor is the *pnp* transistor, to which the same description of operation applies, except that the bias voltages are the opposite way round and the electrons move in the opposite direction. Typical construction, characteristics and symbols for the *npn* transistor are shown in Fig. 7.7. Note the similarity in shape to the pentode characteristics.

The *unipolar* or field effect transistor also makes use of the properties of the *pn* junction, but in a slightly different way. When the junction p.d. is set up across a *pn* junction, the electrons close to the junction in the *n* region move across and fill the holes near the

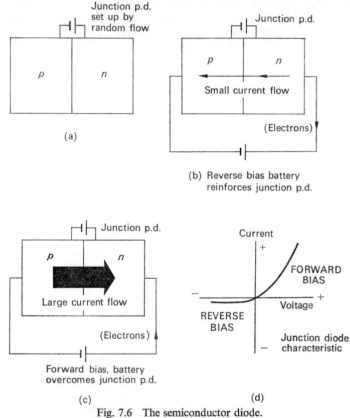

Fig. 7.6   The semiconductor diode.

junction in the *p* region. The area around the junction thus tends to be relatively depleted of available charge carriers and is thus known as a 'depletion zone'. The depletion zone may be extended or reduced by the application of an external p.d. connected so as to aid or oppose the barrier p.d. respectively. In the field-effect transistor, one form of which is illustrated in Fig. 7.8, electrons flow from source to drain, as shown, down the narrow *n*-type channel. The *p*-type gate is held negative with respect to the channel and so the gate-channel junction is reverse biased, setting up a depletion zone whose depth of penetration into the channel is determined by the value of the gate-source reverse bias. Increasing this bias increases the depletion zone and reduces the channel width, thereby reducing current flow for a fixed drain-source p.d. Again, the amplifier condition, that a through current may be changed (modulated) by a small signal, is set up. Typical characteristics of this type of transistor, again similar

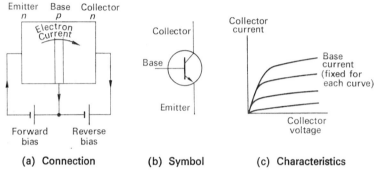

(a) Connection   (b) Symbol   (c) Characteristics

Fig. 7.7   The bipolar transistor.

to the pentode, are shown in Fig. 7.8. In many respects, the field-effect transistor is more like the valve than the bipolar transistor since the FET may be considered voltage controlled. Note than there is an alternative having a p-type channel and an n-type gate. The theory applies equally as before.

The FET described above is of the form known as a junction-gate FET, or JUGFET. A second form in which the input voltage is capacitively coupled via a metal–oxide–silicon arrangement is shown in Fig. 7.9. This type is abbreviated a MOSFET and the applied voltage between gate and source may be used to reduce a channel

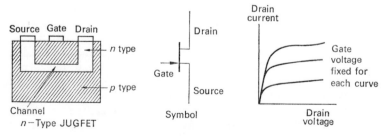

Fig. 7.8   The field effect transistor (JUGFET).

(depletion MOSFET) or create a channel by induction (enhancement MOSFET). An alternative abbreviation is IGFET, which stands for *insulated-gate* FET. A fundamental difference is that in the absence of a signal maximum current flows in the depletion type, whereas no current flows in the enhancement type. For a more detailed discussion of these and other electronic devices the reader is referred to the many specialist electronics texts available.

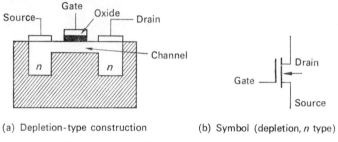

(a) Depletion-type construction   (b) Symbol (depletion, *n* type)

Fig. 7.9   The field effect transistor (IGFET).

7.4   BASIC PRINCIPLES OF ELECTRONIC AMPLIFIERS

To utilise the ability of a valve or transistor which allows through-current variations by the application of small signals, the through current must be fed into a suitable load. This load may simply be resistive as in the case of audio amplifiers, or may be a composite impedance as in radio frequency amplifiers. To illustrate the basic process a simple one-stage audio amplifier using a bipolar transistor will be considered.

The basic circuit is shown in Fig. 7.10. The through current is passed through a resistive load R_L so that the p.d. developed across it will follow the variation in current produced by the signal. In such a circuit the output would normally be taken between collector and ground. If the forward bias is increased by an input signal between base and ground, the collector current rises. This in turn increases the p.d. across R_L, and since the p.d. across R_L plus the p.d. between collector and ground (the output) is constant and equal to the supply voltage, as the load p.d. rises the collector–ground p.d. falls. Thus, the output p.d. follows the input but, in this particular case, with a 180° phase shift. The fidelity of reproduction of the input waveform at the output depends upon the transfer curve linearity as explained above. The amplifier mode illustrated is known as common emitter; other modes are described in Section 7.5.

An important feature of an amplifier circuit is the degree of direct current bias set up in the absence of a signal. To consider again the basic common emitter circuit described above, if when no signal is present the base-emitter bias is zero, then on the application of a signal the transistor would conduct on the negative excursions of the input but would remain cut off on the positive excursions. If it is desired that the output signal should follow both positive and negative excursions then the transistor must be held on in the

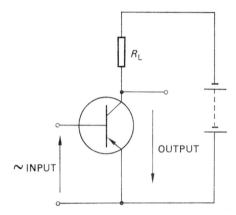

Fig. 7.10   Basic amplifier (common emitter).

absence of a signal. Then when a signal is applied the amplifier would conduct less on positive excursions and more on negative excursions, *i.e.* it would conduct at all times. The degree of bias required is determined by the function of the amplifier as a whole. The bias circuits for different amplifiers are shown in the figures in Section 7.6, which briefly describes more complete circuits.

7.5   ELECTRONIC AMPLIFIER MODES

In all the electronic devices discussed above, the fundamental characteristic rendering them suitable for use in amplifiers is that the device through current is controlled by the application of relatively small input signals. For a valve these signals may be applied either at the grid whilst the cathode is held at a constant potential, or at the cathode whilst the grid is held at constant potential. Similarly, with a transistor, the signal may be applied at the base (or gate) or at the emitter (or source).

The output may be taken from the anode or cathode in the case of a valve or from the collector (drain) or emitter (source) in the case of a transistor.

This set of possible combinations may appear confusing at first. However, they may be summarised as follows: in all cases there

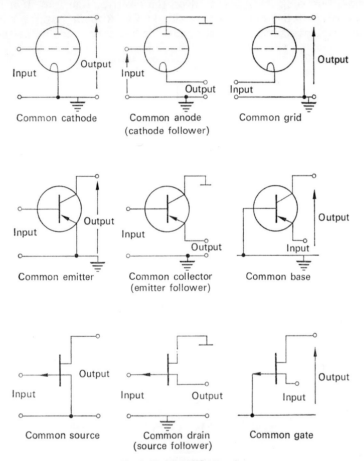

Fig. 7.11 Amplifier modes.

are three electrodes, one of which is the input, one the output and the third is held at constant dc potential or zero ac potential, *i.e.* is effectively grounded to signal. All possible modes are shown in Fig. 7.11. The electrode which is grounded to ac determines the name given to the amplifier mode. Thus, from the figure, there is the common emitter (CE) mode, in which the input is to the base and output from the collector, the emitter being common to both input and output circuits. Similarly, there is the common cathode valve circuit or common source FET circuit, and so on.

Topologically, the common emitter, common cathode and common source circuits are the same as are the common grid, common base and common gate, and the common anode, common collector and common drain. An alternative set of names for the latter is the cathode follower, emitter follower and source follower. Relative values of input and output impedances, voltage and current gains are also similar for these circuits. A summary follows:

	Common electrode		
	Cathode Emitter Source	Grid Base Gate	Anode Collector Drain
Voltage gain	high	high	1
Current gain	high*	1*	high*
Input impedance	medium-low	low	high
Output impedance	medium-high	high	low
Phase shift	180°	180°	zero

* Current gain is not applicable to valve or FET amplifier circuits.

Actual figures are not given since they vary widely depending on the device used.

As can be seen the common grid and common anode groups have opposite characteristics. The most widely used group is the common cathode group which offers a compromise between the extremes of the other two.

7.6 SOME TYPICAL ELECTRONIC AMPLIFIER CIRCUITS

Figure 7.12(*a*) shows a two-stage transistor audio amplifier feeding a loudspeaker. Each stage is a common emitter amplifier. In this circuit R_1, R_2 and R_4 provide the bias network for transistor TR1 with R_3 as the load. C_1 is included to prevent the emitter bias voltage across R_4 from fluctuating at signal frequency. The first stage is coupled to the second via capacitor C_2 which provides a low reactance at signal frequency but prevents the direct voltage at the collector of TR1 from being applied to the base of TR2. R_5, R_6

and R_7 form the bias network for TR2, and the load in this stage is the output transformer feeding the speaker. Capacitor C_3 performs the same function for TR2 as does C_1 for TR1. The output transformer is used to match the output impedance of the second stage, which is a power amplifier, to the speaker input impedance.

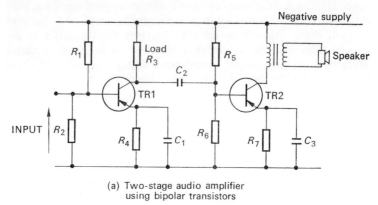

(a) Two-stage audio amplifier
using bipolar transistors

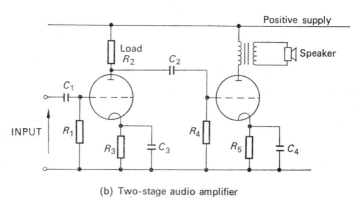

(b) Two-stage audio amplifier
using triode valves

Fig. 7.12 Amplifier circuits.

Figure 7.12(b) shows a circuit performing the same function as that of Fig. 7.12(a), but in this case using triode valves. Bias for each stage is provided by R_3 and R_5 respectively, fluctuation being prevented by C_3 and C_4. C_1 and C_2 are the coupling capacitors and R_2 the first stage load. Note that in this case a current path for the input electrodes is not required since valves are voltage driven devices. The remaining resistors, R_4 and R_1 are called grid leak resistors and are used to prevent the grids from charging to too negative a voltage.

Figure 7.13 shows a single stage pentode radio frequency amplifier. Input transformer T1 and capacitor C_1 form the input tuned circuit which is adjusted to accept a narrow band of frequencies about the centre or resonant frequency. The signal is now passed via V1 to the anode, which again is tuned to provide maximum impedance at the desired frequency using capacitor C_4. R_2 and C_3 provide steady bias for the valve. The screen grid (next to the input grid) is held at a positive voltage equal to the h.t. supply less the p.d. across R_1, the screen dropping resistor. The screen voltage is held steady by capacitor C_2.

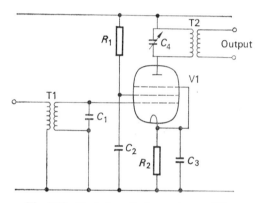

Fig. 7.13   Pentode radio frequency amplifier.

Many other circuits are, of course, possible using a variety of connections as described and using various techniques to improve coupling and to maintain high gain over the range of frequencies and conditions as dictated by the desired function of the amplifier.

7.7   THE MAGNETIC AMPLIFIER

The magnetic amplifier is unlike its electronic counterpart in that the output waveform is not a reproduction of the controlling input. Nevertheless its ability to control high levels of power by the application of small signals allows the device to be included under the general heading. Essentially the magnetic amplifier is an accurately controlled switch having an operating point which is continuously variable.

To facilitate understanding of the basic principle it is necessary to revise the topic of self-inductance and its dependence upon the permeability of the coil core. Self-inductance is the characteristic of a conductor which sets up an induced e.m.f. to oppose any change in coil current. The changing current sets up a changing magnetic flux which in turn induces an e.m.f. according to Faraday's law. The opposing nature of the e.m.f. is described by Lenz's law. Clearly, the magnitude of the opposing e.m.f. is dependent on the magnitude of the flux and this in turn depends upon the current and the magnetic properties of the coil core. The relationship between coil flux and coil current is illustrated by the B/H curve, since B (flux density) depends on flux, and H (magnetising force) depends on current.

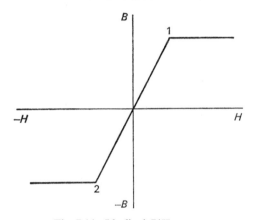

Fig. 7.14   Idealised B/H curve.

Consider now the B/H characteristic illustrated in Fig. 7.14. From the origin to point 1 the flux density increases linearly with magnetising force and so B/H and thus the permeability is constant. The inductance of a core operated in this condition would also be constant. This also applies to the portion of the curve from the origin to point 2. From point 1 increasing H has no effect on B since the core is saturated. Thus B/H and the permeability is zero. Hence the inductance of a coil whose core is saturated would also be zero. Again this applies to the portion of the curve to the left of point 2. The magnetic amplifier employs this principle of core saturation producing zero inductance. The curve of Fig. 7.14 is, of course, idealised but modern core materials can have B/H characteristics which closely approximate the ideal.

The circuit of Fig. 7.15 shows a core having two windings connected so that one winding is in series with a load and an ac supply, and the other winding is provided with a separate dc supply. The device is called a saturable reactor, the windings are termed control and load windings as shown in the figure.

Neglecting the control winding for the moment, provided that the load resistance is much less than the load winding impedance in the unsaturated condition, and also that the ac supply is insufficient by itself to drive the core into saturation, then the bulk of the alternating voltage in the load circuit will appear across the load winding. If now a steady current is allowed to flow in the control winding, then during the half cycles when the load current is in a direction so as to assist the control current, it is possible for the core

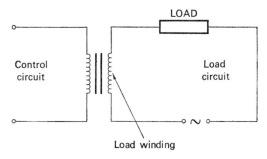

Fig. 7.15   Saturable reactor.

to saturate over part of the half cycle. Once saturated the coil offers zero inductance and during this period the bulk of the alternating voltage appears across the load. Thus the load winding behaves as a switch which is open when the core is not saturated and closed when the core is saturated. The device as indicated is not practicable because of transformer action between the two windings.

Figure 7.16(a) shows two saturable reactors connected so that the control current sets up opposing fluxes in the two cores. This results in one core saturating during part of one half cycle and the other core saturating during an equal part of the other half cycle. This connection also neutralises the induced voltages in the control winding due to transformer action. The load and supply voltage waveforms are shown in Fig. 7.16(b). When one core is saturated its reactance falls to zero, presenting a virtual short circuit to the other reactor which then behaves as a short circuited transformer, so that in turn its reactance is effectively zero. Thus the supply voltage is transferred to the load as in the case of the single reactor. A graph of the average load current against control current is shown in Fig. 7.16(c). This is the transfer characteristic of the amplifier as a whole.

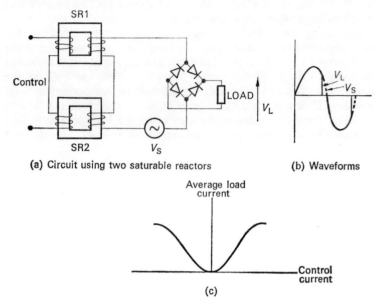

(a) Circuit using two saturable reactors   (b) Waveforms

Fig. 7.16   Transfer characteristic.

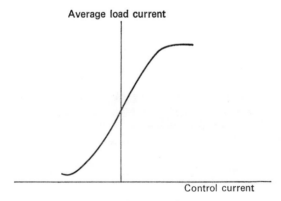

Fig. 7.17   Biased transfer curve.

This characteristic may be shifted to left or right by the addition of bias windings carrying a fixed direct current, the control winding now being the signal input winding (*see* Fig. 7.17).

By feeding back part of the output in a manner such that the feedback aids the input (positive feedback), the amplifier gain may be increased. The necessary windings are called *self-excitation* windings and a simplified circuit is shown in Fig. 7.18. The reader is referred to specialist texts for more detailed discussion of feedback methods.

The main advantages of magnetic amplifiers are robustness, absence of heater supplies and zero 'warm-up' time.

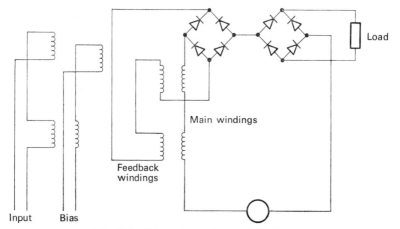

Fig. 7.18   Magnetic amplifier with feedback.

7.8   ROTATING AMPLIFIERS

In a separately excited dc generator, the field current is very much smaller than the armature current, *i.e.* the current provided by the generator for a load. Thus the electrical power input controlling the machine is smaller than the electrical power output, and the machine may be considered an amplifier. (As with all amplifiers the total power output is in fact less than the total power input, the ratio depending on the efficiency. For the purposes of amplification, however, we are considering only one type of power, in this case, electrical. The remainder of the power input is the mechanical power from the prime mover driving the generator.) Most rotating amplifiers are based on the principle of a dc generator having separate excitation.

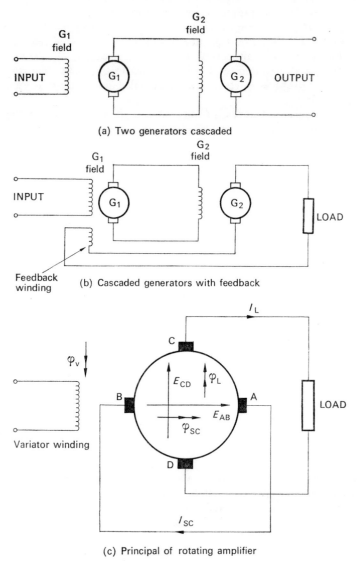

(a) Two generators cascaded

(b) Cascaded generators with feedback

(c) Principal of rotating amplifier

Figure 7.19

If two generators are cascaded as shown in Fig. 7.19(a) the gain of the system is very much increased. However, the response time is also increased since the system inductance is very much larger. If, now, feedback between output and input is introduced by having an additional field winding supplied by the output and connected so that its m.m.f. opposes the main m.m.f., the *rate* of change of the main flux is reduced and so is the effect of the back e.m.f. induced by the changing flux. The response time is thus reduced again (*see* Fig. 7.19(b)).

The system of Fig. 7.19(b), in which an input current and flux controls an output voltage, there being controllable feedback between output and input, may be contained in a single machine shown simply in Fig. 7.19(c).

This type of machine, sometimes referred to as a crossed-field machine, has two sets of brushes, the axes of which are set at right angles to each other. The main field winding is called the variator. In the machine shown the variator flux Φ_v produced by the input current produces an output voltage E_{AB} between brushes AB as shown. The brushes AB are short circuited so that a current I_{SC} flows between them. This current sets up a flux Φ_{SC} which, from basic dc machine theory (*see* Chapter 6), acts along the BA axis. A second voltage E_{CD} is now induced acting along the axis DC. The brushes C and D are connected to the load and a load current I_L flows as shown. The flux set up by I_L, denoted by Φ_L, then acts in opposition to the variator flux as indicated. The flux Φ_L provides the inherent feedback which serves to reduce the response time of the system. The short circuit current is not high because of this inherent feedback. The voltage E_{CD} is proportional to the flux Φ_{SC}.

Φ_{SC} is proportional to I_{SC} provided the value of Φ_{SC} is below saturation and careful design has produced an almost linear relationship. Since brushes AB are short circuited the terminal voltage V_{AB} is zero and since

$$V_{AB} = E_{AB} - I_{SC}R_{AB}$$

where R_{AB} is the armature resistance between A and B and $I_{SC}R_{AB}$ is the armature voltage drop, then

$$E_{AB} - I_{SC}R_{AB} = 0$$

and

$$I_{SC} \text{ is proportional to } E_{AB}$$

Finally, since E_{AB} is proportional to Φ_v, the variator flux, then the output generated voltage E_{CD} is proportional to the variator flux Φ_v and a reasonably linear transfer curve is obtained.

The machine illustrated is known as a *metadyne*. If an additional winding, called a compensator, is introduced in series with the feedback winding so that the compensator m.m.f. opposes the feedback m.m.f., any level of feedback may be obtained. A machine with complete compensation of feedback is known as an *amplidyne*. It follows that an amplidyne has a greater response time than a metadyne; however, this is somewhat compensated for by the fact that as the feedback is reduced the overall gain is increased.

Rotating amplifiers can have very large power gains exceeding 10^5 and are extensively used in the control of heavy rotating or moving loads. In these cases the rotating amplifiers are themselves controlled by electronic or magnetic amplifiers and in this way the advantages of both types of amplifier may be realised.

PROBLEMS ON CHAPTER SEVEN

(1) Describe what is meant by the term 'amplifier'. Categorise amplifiers by function and briefly describe each category.

(2) Define and discuss the following terms applicable to amplifying systems:

(a) gain,

(b) imput impedance,

(c) transfer characteristic.

What particular advantage is gained in using an amplifier with a linear transfer characteristic?

(3) Draw the circuit diagram of a typical valve amplifier for radio frequency amplification and briefly explain the function of each component.

(4) Describe the action of the magnetic amplifier in its simplest form and list the advantages of this device compared to a valve amplifier.

(5) Explain the term 'matching' when applied to electrical circuits. How is matching achieved in electronic amplifiers? The output impedance of a certain valve amplifier is 10 000 Ω. It is to feed a 5 Ω loudspeaker. Describe one method of achieving optimum matching and give precise details of any additional circuit component.

(6) Discuss the advantages and disadvantages of rotating amplifiers compared to alternative systems to perform the same function.

CHAPTER EIGHT

Electrical Instruments

8.1 INTRODUCTION

Electrical indicating and measuring instruments may be broadly divided into two kinds, electromechanical and electronic. Electromechanical instruments are those in which a pointer moves along a calibrated scale (or a scale moves relative to a stationary pointer), the deflection being produced by and proportional to the quantity to be measured. Electromechanical instruments include ammeters, voltmeters, ohmmeters and wattmeters, among others. Electronic instruments are those in which the magnitude of the measured quantity is displayed using electron tubes or semiconductor display devices. Electronic instruments include the cathode ray tube and the digital voltmeter. A valve or transistor volt–ohm meter strictly should be regarded as an electromechanical instrument since here the electronic circuitry is used to improve accuracy rather than for display.

8.2 ELECTROMECHANICAL INSTRUMENT ESSENTIALS

In this type of instrument the moving part is deflected by a force set up by interaction between either magnetic or electric fields which are produced by the quantity to be measured. The moving part is also subject to a controlling force which depends upon the deflection from the zero or null position. When the deflecting and controlling forces are equal the moving part stops and the value of the measured quantity can be read off the calibrated scale. A third essential is a damping force without which the moving part may oscillate about the deflected position. Controlling and damping systems which are to a great extent common to all electromechanical instruments are dealt with in the next section. The nature of the deflection system which further classifies these instruments is dealt with in succeeding sections.

8.3   CONTROL AND DAMPING SYSTEMS

By far the most common form of control system uses hairsprings, either one or two, as shown in Fig. 8.1. The springs are made of phosphor-bronze and have a fairly large number of turns so that the stress produced on deflection is small. When two springs are used they are *contrawound* so that one spring is extended and the other compressed in use; this compensates for any length variation due to ambient temperature change. In the absence of a deflecting force the moving part is held in the zero or null position. In certain instruments, such as the 'Megger', in which springs are not employed as the controlling element, the pointer 'floats' when the instrument is not in use (*see* Section 8.10).

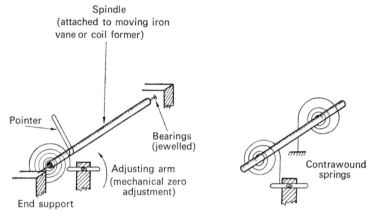

Fig. 8.1   Spring control.

In contrast to electronic indicating instruments the moving part of an electromechanical instrument has inertia and, consequently, on deflection may overshoot the deflected position. To avoid this the moving system is damped either by viscous damping or by eddy current damping. Viscous damping is achieved by attaching to the moving spindle a vane or piston which must displace either air or a suitable fluid when the spindle moves. These systems are shown in Fig. 8.2(*a*). Eddy current damping is an application of Faraday's law of electromagnetic induction. Small currents are induced in metal parts moving through a magnetic field and their direction is such that their associated field opposes the main field, thereby slowing down the movement. In certain instruments such as the permanent-magnet moving coil, the currents induced in the coil former are sufficient for effective damping. In others a special damping disc and extra magnet are included, as shown in Fig. 8.2(*b*).

In all cases it is essential to neither under nor over damp the moving system. The graph shown in Fig. 8.2(*c*) illustrates the position of the moving part from rest plotted against time. Curve (i) shows an

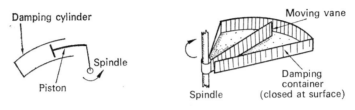

(a) Viscous damping systems

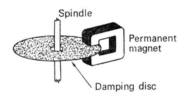

(b) Eddy current damping
using disc and magnet

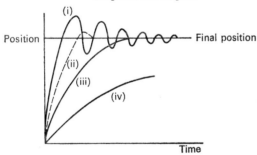

(c) Damping curves

Figure 8.2

underdamped system in which oscillation takes place, curve (iv) an overdamped system which takes too long to settle and curve (iii) a critically damped or *dead beat* system which does not overshoot and comes to rest in a sufficiently short period of time. If for any reason mechanical damage leads to the moving part 'sticking', its behaviour is similar to that shown in (iii) and for that reason curve (ii), in which a slight overshoot is allowed, is usually considered the best one.

8.4 PERMANENT-MAGNET MOVING COIL INSTRUMENTS

The names of the instruments discussed in this and subsequent sections refer to the type of deflecting system employed. One construction using the permanent-magnet moving coil system is illustrated in Fig. 8.3. Another name frequently applied to this type is

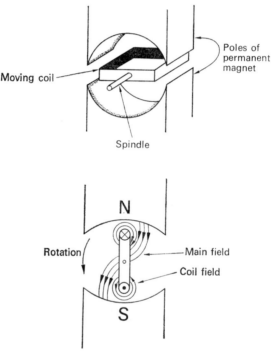

Fig. 8.3   p.m.m.c. (d'Arsonval) movement.

d'Arsonval. As can be seen a coil is suspended so that it can rotate between the poles of a permanent magnet. When current is passed through the coil the magnetic field which is set up reacts with the main field and the coil turns. The controlling system invariably employs hairsprings, as described above, through which the current is led to and from the deflecting coil. For such a system the controlling torque is directly proportional to the angle of deflection, whilst the deflecting torque is directly proportional to the coil current. When the torques are equal, *i.e.* at the deflection position, the angle of deflection is thus directly proportional to the coil current. The scale calibration marks are then equally displaced, *i.e.* the scale is linear.

Such instruments may be employed as ammeters, voltmeters or ohmmeters and, by using a centre-zero position and a coil with many turns, as a sensitive centre-zero galvanometer. Since the direction of action of the deflecting force depends upon current direction, direct application of alternating current at frequencies above a few cycles per second will result in the pointer vibrating at the zero position. If the instrument is to be used with ac supplies, therefore, it is essential to include a rectifying circuit. This is further discussed below.

For use as an ammeter to measure currents above the full scale deflection (f.s.d.) current, a shunt is provided for the excess current. For use as a voltmeter to measure voltages in excess of that which produces the f.s.d. current, a series resistor or multiplier is provided across which is developed the additional p.d. (*see* Fig. 8.4). The use of shunts and multipliers is further discussed in Section 8.8.

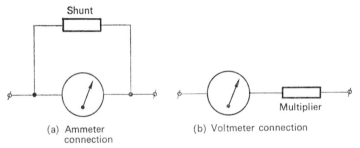

Fig. 8.4   Connection of d'Arsonval instrument as ammeter and voltmeter.

The main advantages of this type of instrument are:

(i) linear scale,
(ii) extension of range is relatively simple,
(iii) inherent damping by eddy currents in the coil former,
(iv) low power consumption.

A disadvantage is that additional circuitry is required for measurement of alternating current. Errors of this instrument and others are discussed in Section 8.9.

8.5 MOVING IRON INSTRUMENTS

In the moving iron type of instrument the deflecting force is again produced electromagnetically, but in this case the coil is stationary and the pointer is attached to a piece of suitably shaped iron which is free to move in the vicinity of the coil. There are two types of moving iron instrument, the attraction type and the repulsion type.

In the attraction type, one form of which is shown in Fig. 8.5(a), an iron vane attached to the pointer spindle is pulled towards the centre of the coil when current flows in the coil. The vane is unmagnetised and so the attraction is in the same direction regardless of

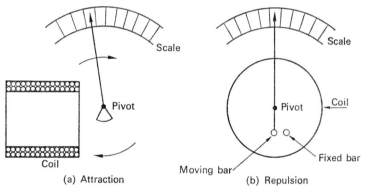

(a) Attraction   (b) Repulsion

Fig. 8.5   Moving iron instruments.

the direction of current flow. (The same principle is used in door chime mechanisms for use with batteries or bell transformers.) In the repulsion type, one form of which is shown in Fig. 8.5(b), there are two pieces of iron, one stationary, one mobile, situated within the fixed coil. When current flows through the coil and a magnetic field is set up, both pieces of iron are similarly magnetised, regardless of current direction, and a repulsive force is set up which pushes the moving iron away from the fixed iron. In both types of instrument the deflecting force is proportional to the square of the current flowing in the coil. Modern moving iron instruments use spring control, as described above, producing a control torque proportional to deflection. The deflection position of the pointer is thus proportional to the square of the coil current and so a 'square law' scale results on which the calibration marks are closer together at the low current or voltage end than at the other end. By specially shaping the iron pieces, the non-uniformity of the scale can be improved on that of a pure square law, but the scale of all moving iron instruments is still essentially nonlinear. Damping is usually achieved by a viscous method using air, as described in Section 8.3.

Moving iron instruments may be employed as ammeters or voltmeters, and since they are not sensitive to current direction may be used equally easily in ac or dc circuits. Principally, they are employed as ac instruments, and in this case extension of range is achieved by using instrument transformers; these are more fully described in Section 8.8.

The main advantages of this type of instrument are:

  (i) simplicity of construction,
 (ii) ruggedness,
(iii) relatively cheap to produce.

The main disadvantage is the nonlinear scale. Errors of this instrument and others are discussed in Section 8.9.

8.6 ELECTRODYNAMIC (DYNAMOMETER) INSTRUMENTS

In the d'Arsonval instrument discussed in Section 8.4, the deflecting force is produced by interaction between a magnetic field due to a permanent magnet and the magnetic field set up by a current passing through the moving coil. If the permanent magnet field system is

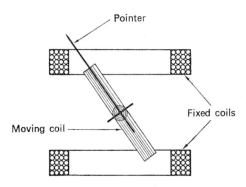

Fig. 8.6 Dynamometer instrument.

replaced by an electromagnet, as shown in Fig. 8.6, an electrodynamic or dynamometer instrument is produced. The instrument thus has two sets of coils, one set being stationary and the other being free to move. Since the deflection is caused by a reaction between two fields, each of which is set up by a current, the deflection is proportional to the product of the two coil currents. Dynamometer instruments may be employed as voltmeters or ammeters as illustrated in Fig. 8.7 in which case the scale would be nonlinear. In practice the most common application is that of a wattmeter which is

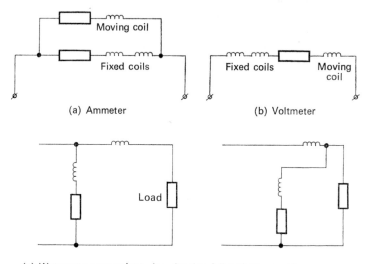

(a) Ammeter   (b) Voltmeter

(c) Wattmeter connections (see Section 8·7 and Chapter 9)

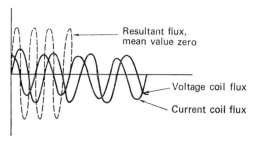

(d) Dynamometer wattmeter in purely apacitive circuit

Fig. 8.7   Use of dynamometer instruments.

described in detail in Section 8.7. If the instrument coils are correctly connected, it can be arranged that the two fields change direction together on reversal of the supply so that the resulting deflecting force is always in the same direction. Consequently, the instrument may be employed in either ac or dc circuits.

Control is usually provided by the hairspring technique and damping may be achieved by either viscous or eddy current means, in the latter case using a disc rotating between the poles of a small permanent magnet.

The main advantage of the instrument lies in its use as a means of measuring power, described in Section 8.7. Errors are discussed in Section 8.9.

8.7 THE DYNAMOMETER WATTMETER

If one set of coils in a dynamometer instrument is connected in parallel with a load, the current in the coils and the resulting field will be determined by the p.d. across the load. If the other set of coils is connected in series with the load, the magnetic field set up by these coils will be determined by the load current (see Fig. 8.7(c)). As was indicated above, the deflection of the moving coil in a dynamometer instrument depends upon the product of the two fields and thus upon the quantities causing them. In a dynamometer ammeter or voltmeter this causes the 'square law' relationship between deflection and measured quantity. In this connection, however, the deflection is determined by the voltage–current product. The connection may be made in ac or dc circuits; in dc circuits the voltage–current product is fixed for a particular circuit and the moving coil deflects accordingly, in ac circuits the voltage–current product is changing even for a fixed circuit, since voltage and current are changing, but the moving coil now takes up an average position and deflects according to the mean value of the voltage–current product. In both types of circuit the instrument may thus be calibrated to read power directly. It should be noted that in dc circuits the power in a load may be obtained by reading voltage and current on separate meters and multiplying the readings together. This does not apply in ac circuits since separate meters take no account of the phase relationship between the voltage and current, and thus of the power factor. A dynamometer instrument, however, does take the phase relationship into account since the one movement of the moving coil is determined by the joint effect of the voltage and current acting together. This may be further illustrated by considering the case where voltage and current act in quadrature (e.g. a loss free capacitor). Separate meters would indicate voltage and current, and the voltage–current product obtained in this manner is not the circuit power. For the dynamometer wattmeter, assuming the current in the voltage coil is in phase with voltage, i.e. neglecting the voltage coil inductance (see below), the two fluxes causing the deflection are in quadrature and their product is as shown in Fig. 8.7(*d*). The mean value of this product is zero and therefore so is the deflection. For all other phase angles between zero and quadrature the mean value of the flux product is not zero and the instrument indicates accordingly.

The construction of a dynamometer wattmeter is as outlined in Section 8.6. Air or iron cores may be used for the fixed coils depending on the purpose for which the meter is intended. If iron cores are used a high permeability material is employed so that the reluctance of the magnetic circuit is composed mainly of the air gap between fixed and moving coils, and thus hysteresis effects are reduced to a minimum. In addition, the magnetic material is laminated since in ac circuits eddy currents will be set up and the magnetic fields associated with them may react with the main coil fields and produce error. Errors inherent in the instrument are discussed in Section 8.9. Use of the instrument in the measurement of power is discussed in Section 8.10.

8.8   EXTENSION OF INSTRUMENT RANGES

Permanent-magnet moving coil instruments

As was discussed in Section 8.4, the basic permanent-magnet moving coil or d'Arsonval instrument may be used directly to measure direct currents or voltages. However, the ranges covered would be extremely small, and in order to obtain useful ranges of current or voltage additional circuitry is required. For use as an ammeter, a shunt resistor is required, and as a voltmeter a series resistor, or multiplier, is required. Connection of these resistors is shown in Fig. 8.4. The method of calculation of resistor values is demonstrated in the following examples.

Example 8.1

Calculate the resistance of the shunt required in a 0–5 A dc ammeter if the basic movement is of the d'Arsonval type having a 15 mA f.s.d. current and resistance 5 Ω.

The p.d. across the instrument for full scale deflection

$$= 5 \times 0.015$$
$$= 0.075 \text{ V}$$

The shunt current for full scale deflection

$$= 5 - 0.015$$
$$= 4.985 \text{ A}$$

The p.d. across the shunt (0·075 V)

$$= 4.985 \times \text{shunt resistance for full scale deflection}$$

Thus shunt resistance

$$= \frac{0.075}{4.985}$$

$$= 0.1505 \ \Omega$$

Example 8.2

A permanent-magnet moving coil meter has a scale reading of 2 mA. The coil resistance is 10 Ω. Calculate the value of the shunt resistance to allow it to read as an ammeter with a scale deflection of 0–2 A.

The f.s.d. p.d. across the instrument

$$= 10 \times 0.002$$
$$= 0.02 \text{ V}$$

The f.s.d. shunt current

$$= 2 - 0.002$$
$$= 1.998 \text{ A}$$

The f.s.d. shunt p.d. (0·02 V)

$$= 1.998 \times \text{shunt resistance}$$

Thus shunt resistance

$$= \frac{0.02}{1.998}$$

$$= 0.0100 \ \Omega$$

Example 8.3

Calculate the value of the multiplier required to use the meter of Example 8.1 as a voltmeter reading 0–10 V dc.

From Example 8.1, f.s.d. p.d. across instrument

$$= 0.075 \text{ V}$$

Thus, at full scale deflection, multiplier p.d.

$$= 10 - 0.075$$
$$= 9.925 \text{ V}$$

The f.s.d. current

$$= 0 \cdot 015 \text{ A}$$

Hence, multiplier resistance

$$= \frac{9 \cdot 925}{0 \cdot 015}$$

$$= 661 \cdot 6 \, \Omega$$

Example 8.4
The d'Arsonval movement of Example 8.2 is to be incorporated into a voltmeter reading 0–100 V dc. Calculate the required multiplier resistance.

From Example 8.2, the f.s.d. p.d. across the movement

$$= 0 \cdot 02 \text{ V}$$

Thus, at full scale deflection, multiplier p.d.

$$= 100 - 0 \cdot 02$$
$$= 99 \cdot 98 \text{ V}$$

The f.s.d. current

$$= 0 \cdot 002 \text{ A}$$

Hence, multiplier resistance

$$= \frac{99 \cdot 98}{0 \cdot 002}$$

$$= 49 \, 990 \, \Omega$$

In practice, both shunts and multipliers are made of material having a low temperature coefficient of resistance, exhibiting a low thermo-electric effect and capable of carrying full scale current without undue rise in temperature.

For use in ac circuits p.m.m.c. instruments, which are sensitive to current direction, require a rectifier to ensure the input current is unidirectional (*see* Fig. 8.8). Errors introduced by this technique and considerations to be made in use are dealt with in Section 8.9 and Chapter 9 respectively. Provided a rectifier is included, shunts and multipliers may be used with ac as well as dc and the calculations are similar. In this case, however, it must be arranged that the additional resistors are non-inductive since, although the meter current is unidirectional, it is fluctuating as shown in Fig. 8.9.

The deflection produced by a p.m.m.c. movement when fed with pulsating unidirectional current as shown in the figure is proportional to the average value of the current. The scale may, however, be calibrated in r.m.s. values, provided the form factor of the input current waveform is known and constant.

The range of ac instruments using the rectifier–p.m.m.c. arrangement may also be extended by the use of instrument transformers as described below and in certain multipurpose instruments a combination of shunts, multipliers and instrument transformers together with the rectifier–p.m.m.c. movement may be encountered.

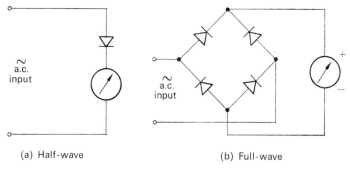

(a) Half-wave   (b) Full-wave

Fig. 8.8   Rectifier–p.m.m.c. instrument.

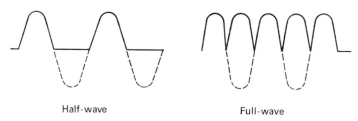

Half-wave   Full-wave

Fig. 8.9   Rectified ac waveforms.

Moving iron instruments

As was stated in Section 8.5, the moving iron instrument is not sensitive to current direction and, unlike the p.m.m.c. instruments, can be employed in ac or dc circuits without additional rectifiers. For dc use the required range may be obtained by adjustment of the number of coil turns and, in addition, by using a suitable series resistor if the instrument is to be employed as a voltmeter. The following example shows a typical calculation.

Example 8.5
To obtain full scale deflection of the pointer in a moving iron instrument, it is found that a total m.m.f. of 1000 A is required. Calculate the number of coil turns if the instrument is to be used as (*a*) a 0–2 A dc ammeter, (*b*) a 25 mA f.s.d. 0–100 V dc voltmeter.

(*a*) The total m.m.f. for f.s.d. is 1000 A, the f.s.d. current is 2 A, thus the number of turns is 500 if the instrument is to be used as an ammeter.

(*b*) The total m.m.f for f.s.d. is 1000 A, the f.s.d. current is 25 mA, thus the number of coil turns required for use as a voltmeter is

$$\frac{1000}{0\cdot025}$$

i.e. 40 000.

In the latter case, the meter would require a total resistance sufficient to reduce the f.s.d. current to 0·025 A when f.s.d. voltage is applied, *i.e.* a resistance of 100/0·025, *i.e.* 4000 Ω.

The additional resistor to be connected in series with the instrument coil would have to have a resistance of 4000 Ω less the resistance of the 40 000 turn coil.

The use of instrument transformers
For ac use the required range of a moving iron instrument is commonly obtained using instrument transformers. These may also be used with p.m.m.c. instruments and to extend the range of dynamometer wattmeters to be used in ac circuits (*see* Fig. 8.10). There are two types of instrument transformer, (i) the current transformer and (ii) the voltage transformer.

(i) It was shown in Chapter 5 that the use of a transformer changes both voltage and current levels in an ac circuit, the nature of change of voltage being opposite to that of current, *i.e.* if voltage is increased from primary to secondary, current is reduced in approximately the same ratio. Current transformers for use with instruments are wound specifically with this in mind and are characterised by a low number of turns on the primary winding and a high number of turns on the secondary. Very commonly in high current circuits the primary winding is, in fact, the supply line whose current is being measured and the secondary winding is mounted on a suitable core which fits over the supply line (*see* Fig. 8.11). Such transformers are called bar-primary transformers as opposed to wound-primary transformers, the latter being used in low-current, high-accuracy applications, *e.g.* laboratory and bench work.

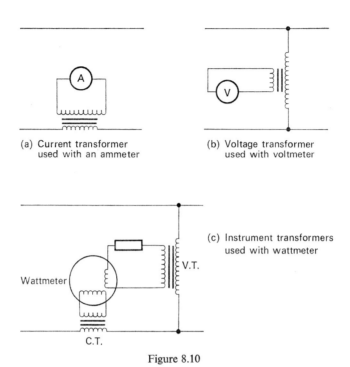

(a) Current transformer
used with an ammeter

(b) Voltage transformer
used with voltmeter

(c) Instrument transformers
used with wattmeter

Figure 8.10

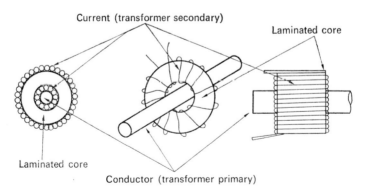

Fig. 8.11 Bar-primary current transformer.

The phasor diagram for a current transformer assuming zero power factor at the load side and neglecting the impedance of both windings is shown in Fig. 8.12. In the phasor diagram shown, a 180° phase shift is assumed in the transformer. It should be noted, however, that, as was pointed out in Chapter 5, the phase shift may be zero or 180° depending upon which point is taken as reference. The 180° phase shift in this case was assumed in order to improve clarity of the phasor diagram. The phasors shown are as follows:

I_p is the primary current,
I_o is the exciting current component of I_p, made up of I_c, which is the core loss component, and I_{mag}, which is the magnetising component,
I_s is the secondary current,
nI_s is the component of I_p providing the necessary ampere-turns for the secondary current I_s.

The constant n is the secondary:primary turns ratio.

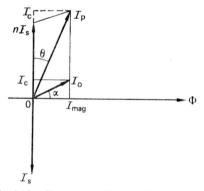

Fig. 8.12   Current transformer phasor diagram.

Two main forms of error are associated with current transformers. These are ratio error and phase angle error. For accurate measurement it is clear that the primary to secondary current ratio should have a constant known value. In fact this is not so.

From the phasor diagram,

$$I_p{}^2 = I_{mag}{}^2 + (nI_s + I_c)^2$$

Hence

$$I_p{}^2 = I_{mag}{}^2 + n^2 I_s{}^2 + 2nI_s I_c + I_c{}^2$$
$$= I_o{}^2 + n^2 I_s{}^2 + 2nI_s I_c \quad \text{since} \quad I_o{}^2 = I_{mag}{}^2 + I_c{}^2$$

If I_o is small and $I_o{}^2$ may be neglected

$$I_p{}^2 \simeq n^2 I_s{}^2 + 2n I_s I_c$$

and

$$\frac{I_p{}^2}{I_s{}^2} \simeq n^2 + 2n \frac{I_c}{I_s} \qquad (8.1)$$

Now

$$\left(n + \frac{I_c}{I_s}\right)^2 = n^2 + 2n\frac{I_c}{I_s} + \left(\frac{I_c}{I_s}\right)^2$$

and since $(I_c/I_s)^2$ will be very small

$$\left(n + \frac{I_c}{I_s}\right)^2 \simeq n^2 + 2n\frac{I_c}{I_s} \qquad (8.2)$$

Comparing eqns. (8.1) and (8.2),

$$\left(\frac{I_p}{I_s}\right)^2 \simeq \left(n + \frac{I_c}{I_s}\right)^2$$

so

$$\frac{I_p}{I_s} \simeq n + \frac{I_c}{I_s} \qquad (8.3)$$

and it is seen that the ratio of I_p to I_s is not dependent only on the turns ratio n, which is constant, but also on the core loss component I_c of the exciting current and also on the secondary current I_s. This error is called *ratio error*. Ratio error may be reduced by the use of a low loss core material such as mumetal, which then reduces I_c in eqn. (8.3), and by increasing the turns ratio n, which makes the I_c/I_s term in the right-hand side of eqn. (8.3) a smaller part of the whole and also increases I_s, which further reduces I_c/I_s. As can be seen from the phasor diagram, ratio error is reduced in high primary current application for then the exciting current and thus the core loss current is a smaller proportion of the primary current. Ratio error should be taken into consideration in all applications of current transformers.

The primary current of a current transformer is mainly the load current being measured and does not change when the secondary winding is open circuited as is the case with 'ordinary' transformers. Consequently on open circuit the large primary current will act entirely as a magnetising current and may produce a dangerously high induced e.m.f. Current transformers should not be open circuited.

The second main form of error, phase angle error, is not important in current measurements, *i.e.* where the transformer is used to extend the range of moving iron or rectifier–p.m.m.c. ammeters, but must be taken into account when the current transformer is used in conjunction with a wattmeter. For this application it is important that the wattmeter current–coil current should be either in phase or exact antiphase with the load current. As can be seen in Fig. 8.12, the primary (load) current is not in exact phase or antiphase with the secondary (current–coil) current but is displaced slightly, again due to the presence of the exciting current I_o. The phase angle of the transformer θ is the angle between the reversed secondary current (lying along nI_s) and the primary current I_p.

From the diagram,

$$\tan \theta = \frac{I_o \cos \alpha}{nI_s + I_o \sin \alpha}$$

where α is the angle between I_o and the flux phasor. If θ is small and expressed in radians

$$\theta \simeq \tan \theta$$

and so

$$\theta \simeq \frac{I_o \cos \alpha}{nI_s + I_o \sin \alpha} \text{ rad}$$

and since $I_o \sin \alpha$ is small compared with nI_s

$$\theta \simeq \frac{I_o \cos \alpha}{nI_s} \text{ rad}$$

From the diagram, $I_o \cos \alpha = I_{mag}$, the magnetising component, and so

$$\theta \simeq \frac{I_{mag}}{nI_s} \text{ rad} \tag{8.4}$$

The phase angle may be reduced by reducing I_{mag} and by increasing the turns ratio and the secondary current. Thus, the steps taken to reduce ratio error will also reduce phase angle error.

(ii) Voltage instrument transformers are used to reduce the level of the measured voltage to a value suitable for the moving iron or rectifier–p.m.m.c. instrument being used as a voltmeter or for the voltage coil of a dynamometer instrument being used as a wattmeter. The theory is the same as that for the normal transformer being operated in an approximately open circuit condition. This is so since the secondary current is very low and the primary current is very nearly equal to the exciting current I_0. The phasor diagram assuming unity power factor on the load side is shown in Fig. 8.13. In this figure

I_p, I_c, I_{mag} are as for Fig. 8.12,
V_p is the primary applied voltage,
V_{pz} is the primary winding voltage drop,
E_p is the component of V_p in opposition to E_s,
V_s is the secondary voltage,
V_{sz} is the secondary winding voltage drop,
E_s is the secondary induced e.m.f.,
θ is the phase angle between V_p and V_s.

As can be seen, in this case the impedance of the windings is taken into consideration. The errors introduced by a voltage instrument transformer are similar to those discussed above. The ratio of V_p to V_s is not the turns ratio but includes a term containing the impedance components of the windings and the exciting current components

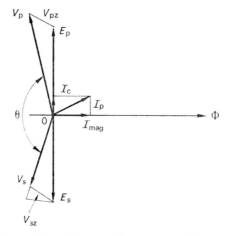

Fig. 8.13 Voltage transformer phasor diagram.

I_c and I_{mag}. This introduces an error particularly important in voltmeter applications. The phase angle θ between V_p and V_s, which introduces an error particularly important in wattmeter applications, can also be shown to be dependent on the windings impedance and on the components of I_0. If the resistance and leakage reactance of the windings are kept as low as possible, the latter by having the windings as close together as possible, and a high permeability core material is used to reduce I_o as much as possible, the ratio and phase angle errors are quite small and in any case are not as serious as those involved when using a current transformer.

8.9 INSTRUMENT ERRORS AND GRADING

In general, instruments are graded according to the accuracy which may be expected of them. The classification and grading is according to the British Standards Institution publication BS 89:1954 from which the following summary is drawn.

There are two grades of instrument, *precision* and *industrial*. Precision instruments are constructed for high accuracy and are used in precise laboratory work and for calibration purposes. The limits of error are laid down as 0·3 per cent of the scale range for p.m.m.c. instruments, and 0·5 per cent of the scale range for all other forms of instrument used in voltage or current measurement, and between 0·5 per cent and 1·0 per cent of the scale range for wattmeter applications, depending on whether the instrument is single or multi-range and on whether the load is balanced or otherwise. In addition, indication must not vary by more than 0·03 per cent to 0·1 per cent if the ambient temperature changes by $\pm1°C$, the exact limit again depending on function. Industrial instruments are used in applications where high accuracy is not required and limits are laid down as follows: up to 2·0 per cent of the scale range for p.m.m.c. instruments and moving iron instruments and up to 2·5 per cent of the scale range for electrodynamic wattmeter applications. The exact figures depend upon function and on the scale length; the reader is referred to the BS publication for precise details.

The error limit per degree Centigrade ambient temperature change lies between 0·1 and 0·2 per cent, and again the exact figure depends on function and type of instrument.

A third instrument type is a 'hybrid' version used for laboratory industrial applications, and its limits lie between those for precision and industrial types.

A number of errors common to all instruments will now be considered followed by a discussion of errors of specific types. The main errors common to all instruments are those due to friction and temperature. Friction errors are due to wear of the system used for suspending the indicating part of the instrument. An important factor is the 'torque to weight' ratio which indicates the magnitude of the deflecting force relative to the weight of the moving part. Friction problems are generally reduced by the use of a vertical spindle rather than a horizontal one and it is important to use instruments such as laboratory industrial and precision types in the position in which they have been calibrated, *i.e.* with *scale* horizontal. Friction losses may be removed by the use of a suspension 'ribbon' to support the moving coil. This provides the control torque as well, thus removing the need for hairsprings. The instrument is quite robust and is called 'pivotless'. Temperature changes increase the resistance of coils and may extend the length of the hairspring where this method of control is used.

Various precautions in the choice of materials, siting of shunts or multipliers and construction of the overall device are taken during manufacture. In use, it is important to allow adequate ventilation of instruments and to avoid use in regions likely to experience large temperature fluctuations. Special transducers are employed for measurements in such regions encountered industrially.

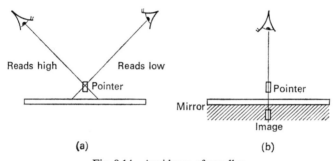

Fig. 8.14 Avoidance of parallax.

A third error common to all instruments in reading the deflection is due to parallax. This is illustrated in Fig. 8.14(*a*) and is due to the position of the observer relative to the pointer and scale. Such errors are reduced in precision and laboratory industrial instruments by the use of a knife-edge pointer and a scale mirror. Care must be taken to 'line up' the pointer and its image as shown in Fig. 8.14(*b*), the reading being that then lying between the two.

Finally, in use all the instruments considered should be suitably shielded from stray magnetic fields, which may be of a magnitude large enough to react with the internal deflection fields.

A summary of specific instrument errors is as follows.

Permanent-magnet moving coil
Well used instruments of this type may suffer from a weakening permanent field which will result in a reduced deflection. Other errors are generally low since the moving coil is well shielded from external

fields, the construction is such that a high torque/weight ratio may be achieved, and, since the instrument generally operates at low power levels, internal variation in temperature is slight.

Alternating current instruments utilising p.m.m.c. movements in conjunction with rectifiers are subject to special errors. The deflection of the instrument is proportional to the average value of the coil current, the scale normally being calibrated in r.m.s. values of current or voltage. This means that the actual reading is equal to the form factor × average value and that if the instrument is used to measure alternating quantities having a waveform other than that used for calibration, error will inevitably result. Certain points arising from this, which should be observed in use, are considered in Chapter 9. One other error of which to be aware is that caused by frequency; this is also considered in the next chapter. Errors incurred when instrument transformers are used with a rectifier–p.m.m.c. instrument were considered in Section 8.8.

Moving iron instruments

The most important error peculiar to the moving iron construction is that due to hysteresis of the magnetic circuit of the instrument. This results in the flux density of the deflecting field for a particular value of coil current being dependent not only on the magnitude of the current but also on whether, in a series of readings, the previous current level to that measured was higher or lower. Choice of magnetic material and careful design of the physical construction of the instrument helps to some extent, but the error, especially in applications involving continuous measurement of the current or voltage, can never be entirely eliminated. In ac applications the effect of the coil impedance variation for different frequency levels must be taken into consideration for work requiring reasonable accuracy. The coil inductance may be suitably swamped by the high series resistance in moving iron ac voltmeters but the error introduced in moving iron ac ammeters is not as easily removed.

Moving iron instruments usually work at higher power levels than p.m.m.c. instruments, and the possibility of temperature variation is rather more important. Errors incurred when instrument transformers are used in conjunction with these instruments were considered in Section 8.8.

Dynamometer instruments

As was stated earlier in the chapter, dynamometer instruments may be air or iron cored. Each variety has its own problems. If a magnetic material is used, errors due to hysteresis and eddy currents are set up. On the other hand if an air core is used the resultant deflecting field is quite small and the moving coil turns must be increased to compensate. This, in turn, reduces the torque to weight ratio, thereby increasing friction losses.

Dynamometer instruments used in ac circuits have an error associated with them caused by the impedance of the coils. This shifts the current and voltage of the voltage coil out of phase. The error is particularly noticeable in wattmeter applications, and for precise work a correction factor is applied as follows.

The voltage coil is inductive and so the coil current, i_v say, lags the applied voltage V by an angle θ depending on the coil impedance. This angle is of course frequency dependent, increasing with supply frequency. For a lagging power factor for which the applied voltage leads the supply current by an angle ϕ, the actual angle between the currents in the two coils will be $(\phi - \theta)$ and the wattmeter deflection is proportional to

$$Ii_v \cos (\phi - \theta) \qquad (8.5)$$

where I is the current in the current coil. Now the impedance of the voltmeter coil Z_v is given by

$$Z_v = \frac{R}{\cos \theta} \qquad \left(\text{since } \cos \theta = \frac{R}{Z_v}\right) \qquad (8.6)$$

where R is the total resistance in the voltage coil circuit and so

$$i_v = (V/R) \cos \theta \qquad (8.7)$$

Hence, from eqn. (8.5),

$$\text{deflection} \propto \frac{IV}{R} \cos \theta \cos (\phi - \theta)$$

If θ were zero, the deflection is

$$\propto \frac{IV}{R} \cos \phi$$

and so the ratio between them is

$$\frac{\cos \phi}{\cos \theta (\cos \phi - \theta)}$$

This is called the *correction factor*, and the actual reading should be multiplied by this factor to give the correct value. It can be seen from eqn. (8.6) that θ may be reduced by making R as large as possible and thus R/Z_v almost unity.

If instrument transformers are used with a dynamometer wattmeter, there is the additional phase angle error of both transformers as well as that due to the voltage coil impedance. These may increase or decrease the total error angle depending upon whether the load power factor is leading or lagging.

8.10   OHMMETERS AND 'MEGGER' INSTRUMENTS

An ohmmeter is an instrument having a scale calibrated in ohms, and calculations, as required in the measurement methods of Chapter 9, are not needed. The circuit of a simple ohmmeter using a p.m.m.c. movement is shown in Fig. 8.15. The battery p.d. V is applied to two resistors R and R_x in series. If i is the meter current

$$i = \frac{V}{R + R_x} \tag{8.8}$$

Provided V is constant, i, and thus the deflection, is inversely proportional to $(R + R_x)$. If R_x, the resistance to be measured, is zero, $i = V/R$, and so the value of R may be fixed to give a value of i equal to the f.s.d. current. The scale is marked 'zero ohms' at this point. If R_x is infinite, $i.e.$ the terminals AB are open circuit, the deflection is zero and the scale may be marked 'infinity'. At various points

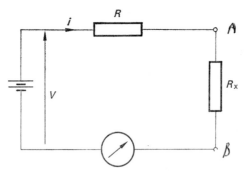

Fig. 8.15   Simple ohmmeter.

between these values the corresponding value of R_x may be marked on the scale during calibration, and if V has the same value the instrument once calibrated may be used to measure ohms directly. In practice the battery runs down with time and so, to compensate, part of R may be made adjustable or a variable shunt resistor across the movement may be added to give a 'set zero' control (*see* Fig. 8.16). In use, the leads of the instrument are connected together and the 'set zero' control adjusted (if necessary) to produce a 'zero ohms' deflection. Direct measurement of the unknown resistance may now be made. The 'set zero' control also compensates for the lead resistance since this is automatically subtracted from the total resistance by adjusting to 'zero ohms' when the leads are in circuit. As can be seen from eqn. (8.8), the i versus R_x relationship is not linear, and so the scale calibrations will not be uniformly spaced. If R_z in the

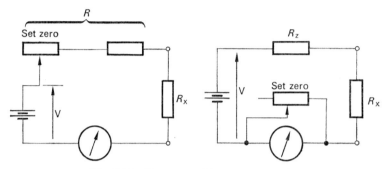

Fig. 8.16   Ohmmeter with set-zero controls.

circuit of Fig. 8.16 is made adjustable in discrete increments, as shown in Fig. 8.17, a multirange ohmmeter may be obtained. The overall accuracy is not high but is usually good enough for laboratory or servicing applications requirements.

The 'Megger' form of ohmmeter which may be also used for continuity and insulation tests of cables (see Chapter 9) is illustrated in Fig. 8.18. In this instrument the pointer is attached to two moving

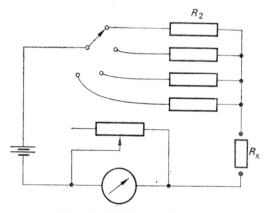

Fig. 8.17   Multirange ohmmeter.

coils which are rigidly fixed at an angle to one another, the whole arrangement being free to rotate between the poles of a bar magnet.

The supply voltage is derived from a hand or motor driven generator and the coils are connected across this so that, when the output circuit is complete and current flows in both coils, the reaction between the main field and potential coil field causes an anticlockwise torque, and the reaction between the main field and the current coil field causes a clockwise torque.

The resultant torque is dependent upon the *ratio* of the potential coil current to the current coil current, and since both coils derive their current from the same voltage, that across the unknown

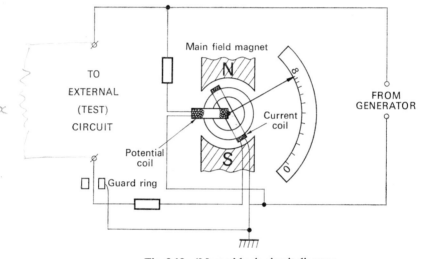

Fig. 8.18 'Megger' basic circuit diagram.

resistor, this voltage does not appear in the ratio equation and the deflection is independent of the instrument e.m.f. If the output terminals are open circuited, *i.e.* the unknown resistance is infinite, only the potential coil carries current and the pointer deflects to the 'infinity' position on the scale. If the output terminals are short circuited, *i.e.* the unknown resistance is zero, only the current coil carries current and the pointer deflects to 'zero' on the scale. Any other value of the unknown resistance will give an intermediate deflection, and the scale may be marked accordingly to read resistance directly. The instrument does not have an additional 'control' mechanism since its action depends upon the mutual reaction between two opposing torques, and thus when not in use the pointer 'floats' indiscriminately over the scale. The guard ring shown in the figure is provided so that leakage currents in the instrument which may occur at the values of test voltage used are shunted to earth and do not affect the reading. The instrument is especially useful in insulation testing under an applied p.d. Methods of use are discussed in Chapter 9.

8.11 ELECTRONIC VOLTMETERS

The use of electronic circuits in association with a p.m.m.c. instrument to give the valve or transistor voltmeter (or volt–ohm meter) can greatly increase sensitivity (*see* Section 9.2) and extend the frequency range.

One useful basic circuit is the 'long tailed pair' or 'difference amplifier' circuit shown in Fig. 8.19(*a*) for valves, and Fig. 8.19(*b*) and (*c*) for transistors. With no p.d. between the input leads, the two valves (or transistors) are equally biased and carry the same level of current in each one. The p.d. between each of the output electrodes, marked A and B in the circuits shown, and ground should then be equal and the p.d. between the electrodes V_{AB} should be zero. A slight difference due to imperfection in the matching of the input valves or transistors may be adjusted by the 'zero adjust' control as shown. The meter, carrying a current dependent on V_{AB}, then reads zero. When a voltage is applied at the input the circuit becomes unbalanced and the left-hand valve or transistor will carry more or less current than the right-hand one. The potential differences between A and ground and between B and ground are now different and V_{AB} is no longer zero. Correct choice of the operating point of the circuit ensures that V_{AB} is proportional to the input voltage and the scale may be calibrated accordingly. A potential divider connected across the input with the input control electrode being connected to a tapping switch, as shown in Fig. 8.20, may be used to give a multi-range instrument.

The sensitivity of the instrument is improved due to two factors; first, the amplification which takes place, and, secondly, the fact that the instrument draws no current in the case of a valve voltmeter and a very small current in the case of a transistor voltmeter. In this respect the field-effect transistor voltmeter is almost as good as the valve circuit since the FET is comparable in many ways to a valve, having a high input impedance and being voltage rather than current operated.

If a circuit similar to that shown in Fig. 8.15 is included at the input and the meter used to measure the p.d. across the unknown resistor, the instrument may be used as an ohmmeter.

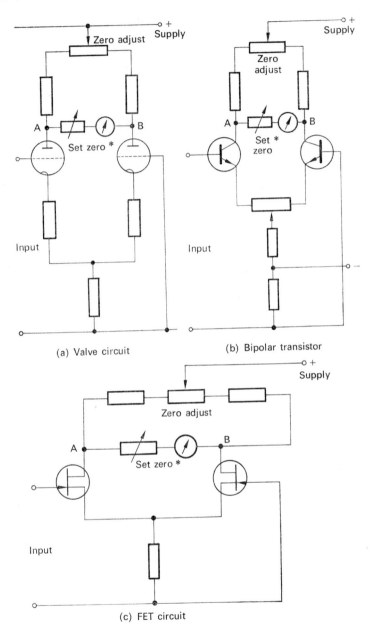

Fig. 8.19   Basic circuit for electronic voltmeter (the set zero* is used only when new valves or transistors are connected; adjust zero is on panel exterior).

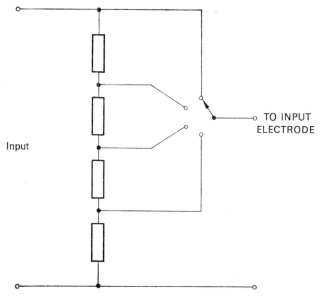

Fig. 8.20 Input potential divider for changing electronic voltmeter scale range.

For ac purposes the input signal is first rectified and filtered, the resultant dc signal then being applied to the input terminals. Sensitivity on ac or dc may be improved by the use of additional amplifier stages.

8.12 ELECTRONIC INSTRUMENTS: THE C.R.O.

Undoubtedly the most versatile of all instruments, the cathode ray oscilloscope is not only able to measure voltage, current, frequency and so on, but a visible picture of the waveform being examined is presented at the same time. Methods of use of the c.r.o. are discussed in more detail in Chapters 9 and 10. In this section the basic construction and operation will be discussed.

The cathode ray tube has an evacuated glass envelope, suitably shaped, containing a means of producing a flow of free electrons, of controlling the flow and of focusing and deflecting the resultant beam. The beam is directed onto a tube face, or screen, which is coated with a material which becomes luminescent under the impact of the electrons. If correctly focused under no signal conditions a fine 'dot' of light appears at the screen.

In detail, the essential parts of a cathode ray tube are:

(a) a cathode similar to that in the thermionic valve,

(b) a modulator electrode or grid which, by means of a suitable electrical potential, controls the beam intensity and thus the brightness of the trace at the screen,

(c) an anode system which is positively biased and accelerates the electrons towards the screen,

(d) a focusing system; this may use electrostatic focusing, in which electric fields control the beam width, or electromagnetic focusing, in which magnetic fields, produced by suitable coils, control the beam width,

(e) a deflecting system; again electrostatic or electromagnetic means of deflection may be employed,

(f) the screen coated with a fluorescent material.

The parts (a), (b) and (c) are collectively called the 'electron gun'. Whether electrostatic or electromagnetic focusing or deflection is employed is determined by the intended use of the tube. All-electrostatic and all-electromagnetic tubes as well as hybrid varieties are available. Tubes used in television sets or radar displays, for example, use electromagnetic deflection since a wider deflection for shorter tube neck is made available with this method. The cathode ray oscilloscope normally employs all-electrostatic tubes, and a cross section of a typical tube is shown in Fig. 8.21. This figure also shows a much simplified control circuit suitable for such a tube.

For the tube shown, electrons are emitted by the heated cathode and are drawn via the three-anode structure (which is positive with respect to the cathode although negative with respect to earth) to the screen. The modulator is negative with respect to the cathode, and by adjustment of its potential the beam intensity and thus brilliance is controlled. The electric fields set up between the three anodes focus the beam as shown in Fig. 8.22, and the fields and thus focal point are adjusted by altering the potential of the middle anode relative to the other two. There are two sets of deflection plates, the x plates, which are used to deflect the spot horizontally, and the y plates, which are used to deflect the spot vertically. The electron beam is attracted towards the more positive of the two plates in each set. For maximum sensitivity to the deflecting voltage at each set of plates the electrons should not be travelling too fast. On the other hand, the brightness obtained at the screen depends upon the energy and thus the speed of the electrons just before impact. Accordingly, the main acceleration takes place after the beam has passed through the deflecting system and this is achieved by the post deflection accelerating anode shown. This is, in fact, a ring of conductive

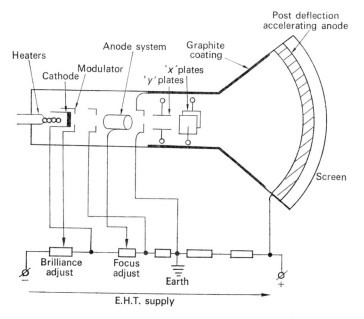

Fig. 8.21 Electrostatic cathode ray tube.

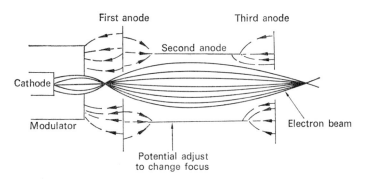

Fig. 8.22 Electrostatic focusing. Electric lines of force go *from* positive to negative, electrons would travel along a line of force in the opposite direction. Focusing occurs due to a combination of this effect and the beam velocity caused by the tube p.d. In the three-anode arrangement the focus anode is 'isolated' from the modulator and brilliance and focus controls do not affect one another.

material around the tube circumference near the screen. The envelope inner surface from screen to third anode is coated with colloidal graphite which effectively acts as a return path from the screen for the electrons. Without this the screen would become negatively charged and tend to repel further electron flow.

The arrangement of the control circuit ensures that the x and y plates are at zero potential with respect to earth whilst still maintaining a high p.d. across the tube from cathode to screen. This ensures safety in operation for the user who will be connecting external circuits to the deflection system.

The use of the c.r.o. in the display of waveforms and measurement of voltage, current and frequency, etc. is considered in the next chapter.

PROBLEMS ON CHAPTER EIGHT

(1) Explain with the aid of diagrams how a moving coil meter may be used to measure (a) a range of currents, (b) a range of voltages, (c) resistance.

(2) A multirange voltmeter calibrated to read 0–50–100–200 V uses a p.m.m.c. movement having a f.s.d. current of 0·015 A and a resistance of 5 Ω. Draw the circuit of the instrument and determine the values of any additional components.

(3) A permanent-magnet moving coil instrument is to be used (a) as a voltmeter reading 0–250 V, (b) as an ammeter reading 0–5 A. The basic movement has a resistance of 10 Ω and requires 10 mA for full scale deflection. Calculate the value of the multiplier for (a) and the shunt for (b).

(4) Describe the principle of action of the permanent-magnet moving coil instrument and the moving iron instrument, and compare advantages and disadvantages of these instruments.

(5) Sketch and explain the construction and action of a dynamometer type wattmeter. What are the causes of error in such an instrument and how may they be reduced?

(6) With the aid of sketches explain how instrument transformers may be used to extend the range of a dynamometer wattmeter. A wattmeter is connected via an 11 kV/110 V voltage transformer and a 200 A/5 A current transformer to measure power in a higher voltage circuit. Calculate the power in the circuit when the wattmeter reads 200 W.

(7) Discuss the use of a current transformer to extend the range of an ac ammeter; include in the discussion the possible errors incurred in such an arrangement and how they may be reduced.

(8) What is meant by the terms 'Industrial Grade' and 'Precision Grade' when applied to electrical instruments? Discuss the errors common to p.m.m.c. and moving iron instruments and how they may be eliminated.

(9) Discuss damping and damping methods employed in electrical instruments. In the discussion define the terms 'underdamped', and 'overdamped' and 'critically damped'.

(10) An alternating current having an average value per half cycle of 2·5 A and an r.m.s. value of 3 A is measured using a rectifier–p.m.m.c. instrument which was calibrated in r.m.s. values using a pure sine wave. Determine the reading on the meter.

(11) Compare the advantages of the cathode ray oscilloscope compared to electromechanical indicating instruments. In a simple sketch illustrate the functional parts of an electrostatic cathode ray tube.

(12) A p.m.m.c. instrument having a resistance of 5 Ω and requiring a f.s.d. current of 15 mA is to be used as a simple ohmmeter with a series set-zero adjustment and a 3 V battery. If the set-zero control is a 0–100 Ω variable resistor, calculate the value of the fixed resistor permanently in circuit. Sketch the circuit.

(13) What is meant by a 'universal instrument'? Sketch and describe a simple circuit of a voltmeter/ammeter with three ranges available for each function.

(14) Describe the advantages of a valve voltmeter compared with a conventional electromechanical instrument. Briefly describe a basic input circuit which could be used in such a meter.

CHAPTER NINE

Methods of Measurement

9.1 INTRODUCTION

This chapter is primarily concerned with the use and application of the instruments considered in the previous chapter in the measurement of voltage, current, resistance, power, frequency and impedance. In addition to the use of these instruments, other methods will be considered as required. The reader is advised to study Chapter 8 before this chapter if maximum understanding is to be achieved.

9.2 MEASUREMENT OF VOLTAGE

The standard methods of voltage measurement include,

   (i) use of voltmeters,
  (ii) potentiometer methods,
 (iii) use of cathode ray oscilloscope.

(i) Use of voltmeters
A voltmeter provides a quick, reasonably accurate method of measuring voltage with the advantage of a direct, visual readout. It is connected across the voltage to be measured. In the choice of a suitable instrument obvious precautions include ensuring that the meter is of the correct type, *i.e.* ac or dc as required, and that the maximum voltage to be measured does not exceed the full scale deflection voltage on the range used. It is common practice to use multipurpose instruments and in this case it is especially important to check that the function switches are correctly set before use. If the voltage being measured cannot be anticipated, the instrument should be set on maximum range to give the initial indication, then switched down as necessary.

    An important characteristic of a voltmeter is the sensitivity, which is usually expressed in *ohms per volt*. Multiplication of this figure by the maximum or range voltage gives the total meter resistance. This determines the accuracy of the reading to a considerable extent since an ideal voltmeter should draw zero current and have infinite resistance. This is illustrated in the following example. Consider the circuit of Fig. 9.1 in which a voltmeter is used to measure the p.d. between points X and Y. Calculation shows this to be one half the applied voltage, *i.e.* 25 V. If the meter used had a sensitivity of 2 kΩ/V (a figure commonly found in cheaper 'pocket' test sets), and the 30 V range were used, the total meter resistance would be 60 kΩ, which is in parallel with the 100 kΩ resistor.

The total resistance between XY is 37·5 kΩ. The p.d. across XY is then 37·5/137·5 × 50, *i.e.* 13·64 V.

If a 20 kΩ/V instrument were used, such as the Avo Model 9, the meter resistance would be 600 kΩ on the 30 V range and the new resistance between XY would be 100 kΩ in parallel with 600 kΩ, giving 85·7 kΩ.

The p.d. across XY is now 85·7/185·7 × 50 *i.e.* 23·1 V.

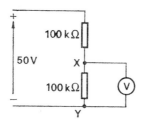

Figure 9.1

The lesson of this example is that the voltmeter resistance should be as high as possible compared to the circuit resistance between the points across which the voltage is to be measured. The ideal voltmeter does not change the voltage level on insertion. This is a particularly important point to bear in mind when measuring alternating voltages, especially in electronic circuits, where the impedance between points is not necessarily that shown in the circuit diagram (due to inherent capacitance, inductance, etc. of devices such as valves or transistors). The ideal voltmeter in these cases, as indeed in most other cases, is the electronic voltmeter which draws no current on insertion.

Another important point to bear in mind when measuring alternating voltages is the type of movement employed. As was shown in the previous chapter, certain movements deflect according to the mean of the square of the quantity being measured whilst others deflect according to the mean value alone. The meter scale in this latter case is invariably calibrated in r.m.s. values on the supposition that the waveform is sinusoidal and the form factor is 1.11. If the waveform is distorted, as it may well be in circuits containing iron cores, the form factor changes and thus an error is introduced into the reading. In cases where this error is suspected and cannot be ignored, it is advisable to check the waveform using a c.r.o., and to find an alternative method of measurement if necessary.

For accurate measurement of small alternating voltages it is not advisable to use rectifier instruments, since the current/voltage characteristic of a rectifier is not perfectly linear, particularly at low values of voltage. Thus non-linearity leads to distortion and resultant change in form factor and contributes an error as described above. The electronic voltmeter or the cathode ray oscilloscope provide suitable alternative methods of measurement in these cases.

(ii) Potentiometer methods

The potentiometer provides a straightforward and extremely accurate method of measuring voltage. Using this fact it can also be applied in turn to measure current and resistance. The principle is illustrated using the simple dc potentiometer arrangement of Fig. 9.2. A battery having an e.m.f. E_1 volts is connected in series with a variable

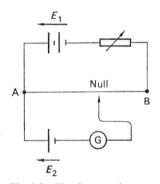

Fig. 9.2 Simple potentiometer.

resistor and a slide wire AB. The slide wire should be of uniform cross section and of homogeneous construction along the working length. If now a second battery of e.m.f. E_2 volts, E_2 being smaller than E_1, is connected via a sensitive centre-zero galvanometer to the slider and one end of the slide wire, polarities as shown, there will be a point along the slide wire at which no current flows in the galvanometer. At this point the voltage between point A and the slider is exactly equal to the open circuit e.m.f. E_2 of the lower battery. At points to the left of the zero or null point, the e.m.f. E_2 exceeds the voltage between A and the slider and the lower battery provides current which indicates in one direction on the galvanometer. At points to the right of the null point, current flows in the galvanometer in the opposite direction, since the voltage between point A and the slider now exceeds E_2. In this case the upper battery is providing the indicating current.

If E_2 is the e.m.f. of a standard cell such as a Weston cadmium cell, for example, then the slide wire voltage drop per unit length is accurately known, *i.e.* if the length A to slider is l_1 metres, the voltage drop per unit length is E_2/l_1 volts per metre. This initial balance point may be adjusted to a convenient place along the wire by alteration of the variable resistance. Once set, however, this resistance is then left unchanged for further measurements. This process of setting and determining the voltage drop per unit length is called *standardising* the potentiometer.

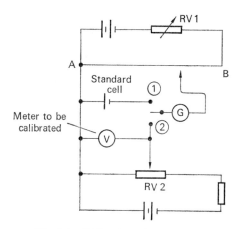

Fig. 9.3   Calibration of a voltmeter.

Once the voltage drop per unit length is accurately known, an unknown voltage may now be connected in the place of battery E_2. The new balance point is determined and the length of slide wire from A to the slider is read off. Suppose this length is l_2 metres, then the unknown voltage is $l_2 \times E_1/l_1$ volts, provided the voltage drop per unit length has been left unchanged. Any alteration of this figure due to run down of battery E_1 or movement of the variable resistor setting will of course introduce error into the measurement. It is advisable when using this method to arrange for null points to be in a position so that the measured slide wire lengths are relatively large. In this way, the effect of non-uniformity of the wire due to end connections is minimised, since the non-uniform portion resistance constitutes only a small portion of the whole. Occasionally, if it is convenient, it is useful to arrange E_2 and l_1 to be numerically equal, e.g. if E_2 is 1·0718 volts, l_1 is arranged to be 1·0718 metres (by adjusting the variable resistor). In this case the unknown voltage will be numerically equal to l_2 and may be read off directly.

Developments of the method in the measurement of current and resistance are considered later in the chapter. The method may be used to calibrate a voltmeter using the circuit of Fig. 9.3. The potentiometer is first calibrated using the switch set to position 1 and then by switching to position 2 the p.d. across the voltmeter may be determined. The voltmeter p.d. may be adjusted by alteration of the variable resistor RV2 and the process repeated to give a series of readings over the entire voltmeter scale.

(iii) Use of cathode ray oscilloscope

As was stated in Chapter 8 the cathode ray oscilloscope is one of the most versatile of all measuring instruments and may be used to indicate the waveform as well as magnitude of any time-varying quantity. Once set up according to the manufacturer's instructions a calibration voltage which is usually made available at the oscilloscope front panel is applied to either the Y or A terminal, and earth and the Y amplitude adjusted to give a convenient waveform size. This control is then left unchanged and the unknown voltage is applied, a comparison then being made by use of the calibrated graticule. The c.r.o. is more commonly used for measurement of fluctuating voltage but direct voltages may also be measured if the trace position for no signal is carefully set and its position noted. On application of a dc signal the trace will deflect vertically in a direction depending on the signal polarity, and the deflection may be read off. For measurement of alternating voltages it is not necessary to worry too much over the timebase frequency if only the magnitude of the voltage is required. The total trace width is then the peak to peak value; this should be borne in mind if comparison is being made between the value as measured by the c.r.o. and that measured by a meter.

Large alternating voltages may be measured using the c.r.o. by trusting the deflection sensitivity (volts/centimetre) to be as marked and not calibrating, since the calibration voltage provided is usually too small; however, for this order of magnitude, meters are probably just as accurate, and the oscilloscope is then reserved for waveform display only.

9.3   MEASUREMENT OF CURRENT

Standard methods of current measurement include,

   (i) use of ammeters,
  (ii) potentiometer methods,
 (iii) use of cathode ray oscilloscope.

Of the three methods, (ii) and (iii) are less commonly used in this context than in voltage measurement.

(i) Use of ammeters

The ammeter is by far the most commonly used method of current measurement. It offers the same advantages from a measurement point of view as the voltmeter, providing a quick, fairly accurate means of measurement with a direct visual readout.

It was stated above that the ideal voltmeter would have infinite resistance and draw no current from the circuit, thereby not affecting any alteration in voltage levels. Similarly, the ideal ammeter should not affect the circuit current distribution which existed prior to insertion. The ammeter, which is connected in series with the circuit part whose current is being measured, should have as *low* a resistance as possible so that minimal p.d. is developed across it. Ammeters do not in general have a 'sensitivity factor' as with voltmeters, but a guide to the quality of the instrument may be obtained from the meter resistance if quoted. In certain applications, for example, transistor circuitry, even the small p.d. which is developed across the instrument may be sufficient to upset efficient working of the circuit, and in these cases it is often advisable to determine current levels by first measuring the voltage across part of the circuit whose resistance is known and then using Ohm's law.

The same precautions as for voltmeters should be observed when the instrument is chosen so that a meter of correct polarity and range is obtained. As before this is particularly important if multipurpose instruments are used and, again, if the approximate current is not known, the initial scale range should be set as high as possible and switched down as required. Incidentally, it is a point worth remembering that range changes of voltmeters and ammeters may reflect into a series of readings in those cases where I/V characteristics of electrical or electronic devices are being obtained.

In the measurement of alternating currents the same considerations described above of the effect of waveform, especially in rectifier instruments, should be borne in mind. Current waveforms may be studied using an oscilloscope and a pure (non-inductive) resistance through which the current is passed to develop the necessary p.d. for application to the deflection system.

(ii) Potentiometer methods

Current may be measured very accurately using potentiometer methods as described in the previous section. The current is passed through a standard resistor whose resistance is accurately known. In order that this resistance shall not affect the circuit in which the measured current is flowing, it should have as low a value as possible which will, of course, result in a small p.d. However, this is quite in order since the potentiometer is very well suited for the measurement of small voltages. The circuit is the standard one shown in Fig. 9.3 with the known resistance connected between point A and point 2.

(iii) Use of cathode ray oscilloscope

The method used here is described in (i) above. The procedure is then the measurement of voltage by c.r.o. as discussed in Section 9.2. As with the potentiometer method the known resistance should be as small as possible, the oscilloscope being quite capable of handling very small voltages.

9.4   MEASUREMENT OF POWER

The easiest and most direct form of power measurement is the wattmeter, the electrodynamic form of which was described in detail in the previous chapter. The instrument may be used in either dc or ac circuits, the range in the latter being suitably extended by the use of instrument transformers.

There will inevitably be an error in the reading of a wattmeter when used as illustrated in Chapter 8. This may be clarified by considering the circuits shown in Fig. 9.4 which illustrate two alternative methods of wattmeter connection.

Figure 9.4(a) shows a wattmeter connected so that the voltage coil lies between the current coil and load. In this case the current flowing in the current coil is not only the load current but also the voltage coil current. The voltage coil current is directly proportional to the voltage so in this case the error is due to the current coil.

Figure 9.4(b) shows a wattmeter connected so that the voltage coil is across both current coil and load. This avoids the problem of the previous circuit since now the current coil carries only the load current. However, the voltage coil current now depends on not only the load voltage but also the current coil p.d. In this case the error is due to the voltage coil.

In both cases the error introduced is small but its importance depends upon the level of power being measured. The error increases with power level and is not constant. If it is required to compensate for error this may be achieved either by calculation (*i.e.* in the circuit of Fig. 9.4(*a*) by deducting the current coil power from the total reading or, in the circuit of Fig. 9.4(*b*), by deducting the voltage coil power from the total reading) or by internal circuit modification using compensating coils so that the coil causing the error also provides a compensating m.m.f. in a direction so as to reduce the total reading by the appropriate amount. The method adopted depends on the particular application and the level of accuracy required.

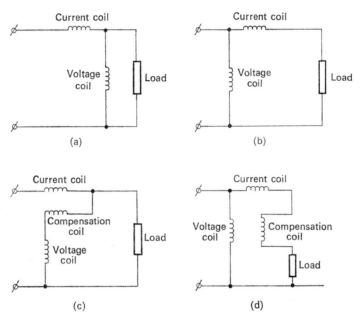

Fig. 9.4   Wattmeter connections.

9.5   MEASUREMENT OF RESISTANCE

There are a considerable number of methods available for the measurement of resistance. The simplest and most convenient method is the resistance or ohmmeter which was described in Chapter 8. Although this method is the most straightforward and offers a direct readout, it is probably the least accurate and is not used where any degree of accuracy is required. The ohmmeter finds application usually when a quick check on a suspect circuit component is required, and since these rarely have a tolerance below

1 per cent, accuracy is not at a premium. Ohmmeters are not generally available as single purpose instruments but usually exist as one of the many modes of operation available of a multipurpose instrument such as the *Avo* or *Selectest*. When using these the instrument is first set to the desired range and the leads are then connected together and the pointer adjusted to zero. This effectively cancels the lead resistance in the instrument reading. The meter should be re-zeroed before use each time and on changing resistance range. Multipurpose instruments should *not* be left switched to the ohmmeter mode when not in use so that the possibility of battery degeneration whilst idle or misuse when next required is avoided.

Continuity testing, which is a form of resistance measurement using the Megger type of instrument, is considered separately in the next section.

One of the most useful non-direct reading methods employed for accurate resistance measurement is based on the Wheatstone bridge principle, which 'balances' the unknown resistance against other known resistances. The basic Wheatstone bridge is shown in Fig. 9.5(*a*). R_x is the unknown resistance, R_A, R_B and R_C being the known resistances. The instrument between points P and R is a sensitive centre-zero galvanometer.

In the circuit shown the source current I_s divides at the node S into two parts, I_C flowing via arm SR and I_x flowing via arm SP. The current flowing in the galvanometer, if any, is due to the differences in potential existing between points P and R. This p.d. depends on the potentials at these points and these, in turn, are determined by the currents I_C and I_x, and the resulting voltage drops along arms SR and SP.

By choosing suitable values of resistance for R_A, R_B and R_C, it can be arranged that the p.d. across arm SR is equal to that across SP so that points R and P are at the same potential with respect to point S and thus no current flows in the galvanometer. This is called the *balance* condition of the bridge. At balance it follows that the p.d. across arm PQ and that across arm RQ are also equal since the same voltage exists across SPQ and SRQ and the voltages across SP and SR are equal. Further, since no current flows in the galvanometer branch of the bridge, the same current, I_x, flows in PQ and SP, and the same current, I_C, flows in SR and RQ.

Since p.d. across PS = p.d. across RS

$$I_x R_x = I_C R_C \qquad (9.1)$$

and since p.d. across PQ = p.d. across QR

$$I_x R_A = I_C R_B \qquad (9.2)$$

Division of eqn. (9.1) by eqn. (9.2) yields

$$\frac{R_x}{R_A} = \frac{R_C}{R_B}$$

so that

$$R_x = R_C \frac{R_A}{R_B} \qquad (9.3)$$

In practice, with the circuit shown, R_A and R_B are preset and are called the *ratio arms* of the bridge. R_C is adjusted until balance is obtained. The ratio R_A and R_B is adjusted so that the required range for R_C for any particular value of unknown resistor is convenient. If R_A/R_B is greater than unity larger values of resistance than R_C can be measured and if smaller than unity smaller values than R_C can be measured. R_A and R_B are usually switchable to 1, 10, 100 or 1000 Ω, and R_C is a decade resistance box giving a maximum value

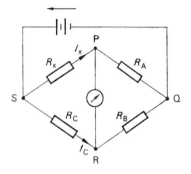

(a) Basic Wheatstone bridge

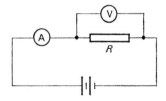
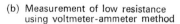

(b) Measurement of low resistance using voltmeter-ammeter method

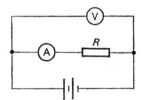

(c) Measurement of high resistance using voltmeter-ammeter method

Fig. 9.5 Resistance measurement.

of resistance of 9999 Ω in 1 Ω steps, or 99 990 Ω in 10 Ω steps. One problem if a low resistance is to be measured is the existence of thermoelectric e.m.f.'s at junctions; the effect of these may be reduced by repeating the measurement with the energy source connections reversed. The measurement is independent of the actual voltage applied, a characteristic which is an advantage. The principle of bridge measurement is an important one and is used as the basis of a number of commercially available instruments, including those for the measurement of reactive quantities. Some ac bridges are described briefly later in this chapter.

Resistance may also be measured indirectly using voltmeters and ammeters. The circuits are illustrated in Fig. 9.5(b) and (c). For resistors of medium value (compared to the voltmeter resistance) either circuit may be used. For low values Fig. 9.5(b) shows the preferred circuit which has the voltmeter connected directly across the unknown resistor R. This arrangement avoids the ammeter p.d., which may be comparable to the unknown resistor p.d., being included in the voltmeter reading. Since the current in R will greatly exceed that in the voltmeter (assuming a current driven movement is being used), the error introduced by the ammeter reading both currents may be ignored. The circuit shown in Fig. 9.5(c) is used for high resistances since the connection shown avoids the voltmeter current, which may be comparable to the resistor current, being included in the ammeter reading. The ammeter p.d. will be very small compared with the load p.d. and its effect on the voltmeter reading may be ignored. In both cases the resistance R is given by

$$R = \frac{\text{voltmeter reading}}{\text{ammeter reading}}$$

the accuracy being dependent upon the quality of the meters being used. The size of the applied voltage is not important provided that the power dissipated in the load does not exceed the wattage rating. A small supply voltage is usually applied because of this factor.

9.6 CONTINUITY TESTING

Insulation and continuity testing of electrical installations is essentially a measurement of resistance, the figure obtained for insulation, *i.e.* between insulation and conductor, being a high value, that obtained for continuity, *i.e.* between ends of a conductor, being a very low value.

The most commonly used instrument is the 'Megger' type manufactured by Evershed and Vignoles, Ltd. This form of instrument was described in Chapter 8. For insulation testing the instrument is used as described with the external circuit (conductor–insulation) connected in series with the deflecting coil and the 'insulation' scale is marked from zero to infinity. The reading is 'good' at the high end of the scale. For continuity testing the test terminals are switched so that the conductor is in parallel with the deflecting coil. If the resistance is low, the bulk of the current flows via the test circuit and the deflecting coil current is low. The pointer then moves to the low current end of the scale. Since this corresponds to the 'infinity' end of the insulation scale, a separate scale is included for continuity and is marked low at this end. If the resistance is high the deflecting coil current increases and the pointer moves accordingly. The higher resistance end of the continuity scale thus corresponds to the zero end of the insulation scale. Since on a continuity test the resistance should be low the 'good' end of the continuity scale is also the 'good' end of the insulation scale.

If accurate measurement is required using the Megger instrument, it is obtainable in 'bridge' form, in which the control coil is supplied from the energy source, the deflecting (current) coil is connected across the bridge and the movement is used as detector.

9.7 FREQUENCY MEASUREMENT

Frequency may be measured directly by frequency meters or indirectly by bridge methods or by the oscilloscope. The basic forms of frequency meter are considered in Appendix I. The bridge method commonly uses the Wien bridge among others, and the layout of this is included in Section 9.8. In this section the oscilloscope method will be described.

The circuit arrangement is shown in Fig. 9.6(a). A known frequency is applied to the horizontal (x) plates of the c.r.o. with the timebase and sync switched 'off'. The unknown frequency is applied to the vertical (y) plates. The resulting patterns produced at the screen, known as Lissajous figures, are determined by the ratio of the two frequencies applied to the deflection plates. Typical figures are shown in Fig. 9.6(b).

To determine the unknown frequency the number of *horizontal* loops in the Lissajous figure is divided by the number of *vertical* loops, and this ratio is then multiplied by the known frequency. The ratios for the figures shown are given in the diagram.

The method is simple and provides dependable results. Measurements may be made at any frequency within the range of the c.r.o. internal amplifiers (given in the manufacturer's specification).

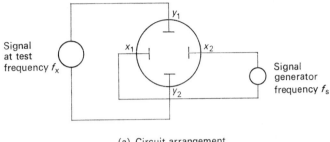

(a) Circuit arrangement

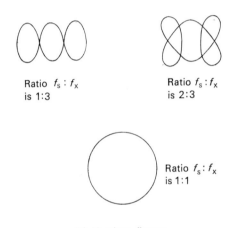

Ratio $f_s : f_x$ is 1:3

Ratio $f_s : f_x$ is 2:3

Ratio $f_s : f_x$ is 1:1

(b) Lissajous. figures

Figure 9.6.

9.8 AC BRIDGES

The Wheatstone bridge principle as described earlier in the measurement of resistance may also be adapted to measure impedance and reactance. Over one hundred different bridge circuits have been suggested for measurement of ac quantities, certain of the more common ones being illustrated in Fig. 9.7. These are described briefly later in the section.

Certain special considerations apply to ac bridges, particularly in the choice of supply and detector. The type of supply is determined by the frequency at which measurements are to be made. At low frequencies the mains supply or variable-frequency generator may be used, at audio frequencies and above an appropriate signal generator together with an amplifier, if necessary, to provide sufficient output is used. The detector may be a microammeter with rectifier, a cathode ray tube (*see* Chapter 8) or at audio frequencies a pair of headphones, again with amplifier if necessary. Care must be taken to match the detector and bridge if maximum sensitivity is to be achieved (*see* Chapter 5). ac bridges are prone to errors due to external influences such as interference, radio frequency noise, etc., and to the effects of internal stray reactances. The effects are largely determined by the frequency at which measurements are made and

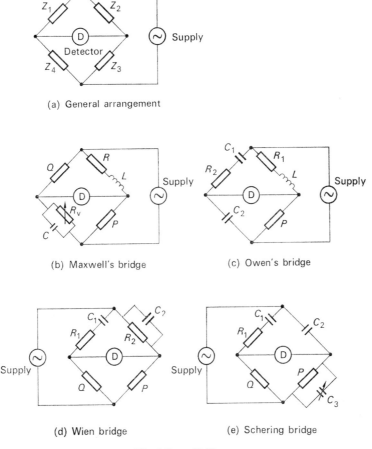

(a) General arrangement

(b) Maxwell's bridge

(c) Owen's bridge

(d) Wien bridge

(e) Schering bridge

Fig. 9.7   ac Bridges.

may be substantially reduced by careful screening of the bridge and the use of accurate components. For very accurate measurement, special circuit configurations are adopted to minimise these various effects. Only the more common bridges are included below, and it should be pointed out that the analysis to give the balance equations is not included. For a more detailed account the reader is referred to the many excellent specialist works on the subject. The general arrangement of an ac bridge is shown in Fig. 9.7(a). For this circuit balance is obtained when

$$\frac{Z_1}{Z_4} = \frac{Z_2}{Z_3}$$

and

$$\theta_1 - \theta_4 = \theta_2 - \theta_3$$

where Z_1, Z_2, Z_3, Z_4 are the impedances shown and θ_1, θ_2, θ_3, θ_4 are the respective phase angles. Note that the impedance balance equation is reminiscent of the Wheatstone equation for the dc case but in addition phase differences between arms must be equal.

Figures 9.7(b), (c), (d) and (e) show the Maxwell bridge, the Owen bridge, the Wien bridge and the Schering bridge respectively.

The Maxwell bridge shown is one of a number of possible configurations. The balance equations are:

$$R = \frac{PQ}{R_v}$$

$$L = PQC$$

The circuit provides a useful means of determination of coil resistance and inductance in one measurement. An alternative circuit is the Owen bridge for which the balance equations are:

$$L = PR_2C_2$$

$$R_1 = \frac{PC_2}{C_1}$$

The Wien bridge is often used for frequency measurement and is also the basis of a well-known electronic oscillator circuit. The balance equations are:

$$\omega^2 = \frac{1}{R_1R_2C_1C_2}$$

so that

$$= \frac{1}{2\pi}\left(\frac{1}{R_1R_2C_1C_2}\right)^{\frac{1}{2}}$$

Finally, the Schering bridge as shown could be used for measurement of capacitance and capacitor loss angle. The equations are:

$$C_1 = \frac{C_2 P}{Q}$$

$$\tan \delta = \omega C_3 P$$

where δ is the loss angle.

In all bridge measurements, standard resistance, capacitance or inductance boxes are used for the fixed arms and accurately adjustable variable resistance or impedance boxes are used for variation to determine balance. The choice of which arms are fixed and which are variable is determined by the particular variable being measured and particularly the possible errors involved in the experimental process.

PROBLEMS ON CHAPTER NINE

The problems in this exercise cover the theory of Chapters 8 and 9.

(1) Explain how a series resistor may be used to extend the range of a moving coil instrument.
A voltmeter having a resistance of 125 000 Ω has full scale deflection at 120 V. When connected in series with an unknown resistor across a 120 V supply the meter indicates 15 V. Find the value of the unknown resistor.

(2) A moving coil meter reads full scale when 3 mA is passed through it. The meter voltage is then 60 mV. Calculate:
(a) the value of the multiplier required to enable the meter to be used as a 0–200 V voltmeter,
(b) the value of the shunt required to enable the meter to be used as a 0–5 A ammeter.

(3) A wattmeter is connected via an 11 kV/110 V voltage transformer and a 150 A/3 A current transformer to measure power in a high voltage circuit. Show the circuit diagram and calculate the power in the high voltage circuit when the wattmeter reads 180 W.

(4) Describe the principle and one use of the Wheatstone bridge.
A Wheatstone bridge consists of four arms, AB, BC, DA which have the following resistances: AB 1000 Ω, BC 100 Ω, DA 22·5 Ω and CD which contains an unknown resistance. Determine the value of the unknown resistance.

(5) Derive an expression relating the ratio arm resistances R_1, R_2 of a Wheatstone bridge with the variable arm resistance R_3 and the

unknown resistance R_x. Assume R_1 and R_x are diametrically opposite to one another. In a circuit as described above $R_1 = 300$, $R_2 = 3000$, $R_3 = 4\cdot21$. Calculate R_x, and hence determine the voltage across this resistor at balance if the bridge voltage supply is 2 V.

(6) Describe a method of measuring resistance using a voltmeter and an ammeter, (a) if the resistance value is high, (b) if the resistance value is low.

A 20 000 ohm per volt voltmeter set on the 0–20 V scale is connected across a 200 kΩ resistor, the combination being connected in series with a 0–1000 A ammeter. Calculate the value of resistance determined by this method if the ammeter reads at the half scale mark. A linear ammeter scale may be assumed. Is this the best method of measurement? Explain your answer.

0–100μA

(7) Describe the two methods of connection of a wattmeter. A wattmeter having a 2000 Ω voltage coil and a 2 Ω current coil is connected with the voltage coil directly across a 500 Ω resistive load. The supply current is 0·1 A. Calculate the wattmeter reading and the actual power in the load.

(8) Describe an instrument containing a facility for continuity testing and normal resistance testing which is independent of supply voltage. Why does such an instrument have a 'floating zero'?

(9) Describe the circuit of a simple ohmmeter and discuss the preliminary adjustments made before use.

A 1 mA f.s.d. moving coil meter of resistance 200 Ω is to be used as a simple series ohmmeter, using a 1·5 V cell. Sketch the circuit showing the value of any additional components necessary.

(10) Explain the essential differences in construction and use of a moving coil voltmeter and an electronic voltmeter using a moving coil movement.

(11) Explain the term 'Lissajous figures' and describe a method of measurement of frequency which uses this phenomenon. In such an experiment a figure containing 12 horizontal loops and 4 vertical loops was obtained when a signal of known frequency of 500 Hz was applied to x-plates of the c.r.o. Calculate the frequency of the signal being measured.

CHAPTER TEN

Testing Methods

10.1 INTRODUCTION: THE NEED FOR TESTING

Many years ago before commerce became the vast and complex system it is today, buying and selling was a relatively simple matter. In market places throughout the world the vendor carried the goods with him, the buyer examined the goods, checked them to ensure they suited his purpose and, if satisfied, made the purchase. The process was simple as befitted the times and the nature of the goods sold. With increasing industrialisation the quantity of goods sold at one time became correspondingly larger and, of course, the goods themselves became more and more complex, necessitating rather more than a cursory examination on purchase. Further, since it became uneconomical to carry around large stocks it became increasingly popular to sell by sample, a method in which sales agreements, often for huge quantities, were established on the basis of a demonstration of a single item. Hence the need to ensure that all items were manufactured to the same specification became more and more important. The need for continuous inspection and testing was thus established.

10.2 THE PURPOSE OF A SPECIFICATION

The dictionary defines 'specify' as 'to describe in detail; state definitely or explicitly'. Thus a specification is a detailed statement of qualities, characteristics, etc. of a particular product. The specification provides a means of communication between manufacturer and customer so that personnel concerned at all stages of production from raw material to finished product know exactly what is required at all times.

The specification defines the materials to be used, the processes involved in manufacture, the finished product, methods of testing and performance criteria for particular methods of use. It will also contain obvious details such as the originator, the manufacturer, the inspectorate (if this is not the originator) and will have some means of identification, probably in the form of an alphanumeric code.

Very often, parts of the specification referring to performance under certain conditions, especially environmental, are common to a wide variety of products, and to this end a considerable number of standard specifications have been evolved by organisations set up for the purpose. These include the British Standards Institution (BSI) of London and the American Standards Association (ASA) of New York. Thus a customer need only refer to the particular standard specifying certain parts of the performance or testing criteria. Similarly, systems of tolerances within which parameters (lengths, diameters, etc.) must lie have also been devised, and again specification detail may be reduced by reference to the standard systems.

10.3 TYPES OF TESTS

The type of testing involved for a particular product is obviously determined by the type of product and its intended use. However, performance tests may generally be divided into two types: performance under normal or rated conditions and performance under adverse conditions. The conditions referred to may also be divided into two types: operational (determined by the product type) and environmental.

Operational conditions depend upon the product. For electrical products, with which we are mainly concerned, these are voltage, current, power and frequency. Testing may be carried out with normal values and with abnormally high or low values of these variables, and observations made as to the limits within which the product performs as specified. Very often it is found that consistency of performance and reliability is improved by operation at values below those for which the product is rated. This process is known as derating.*

Environmental conditions are those of temperature, pressure, humidity, atmospheric content of chemicals and radiation, etc., and again there will be normal or rated values and abnormal values. The effects of both environmental and operational stresses on performance and reliability are discussed in the reference below. The standards of performance of many kinds of products when subjected to various conditions both operational and environmental are laid down by the standards organisations and as such are frequently used as part of a particular specification as described above.

* See *Introduction to Reliability Engineering* by Rhys Lewis, McGraw-Hill, 1970.

10.4   TESTING AND INSPECTION: SAMPLING

To ensure a high standard of quality control so that the specification may be adequately met throughout an entire production run of an item, continuous inspection and testing is necessary. There are three possible ways of carrying out the task; complete inspection of the entire stock, partial inspection based on random sampling and partial inspection based on sampling methods designed on a statistical basis, *i.e.* using the mathematical theory of probability. The first method is expensive, especially for large production runs, and, surprisingly, is not always successful for psychological reasons. The second method being a random method is not reliable, and consequently the majority of inspection techniques are based on the third method.

The subject of statistical analysis and probability theory is large and can be complex. A thorough examination of inspection methods using this theory is thus beyond the scope of this book, and we will confine ourselves to a basic study of vocabulary and a very general discussion of fundamentals. It should be noted that the terms employed and the techniques described are based on those used by the Government Inspectorate. A more detailed account will be found in books listed in the bibliography at the end of the chapter.

Vocabulary of inspection

As with other specialist studies a vocabulary of terms has been established for inspection and testing. The main ones are as follows:

lot or *batch:* a group of items offered for inspection,

defect: a non-conformance of any item to specification,

defective: an item containing one or more defects,

defect per hundred: one hundred times the ratio of total defects in a lot or batch to the number of items in the lot or batch,

per cent defective: the ratio of the number of defective items in a batch to the number of items in the batch, expressed as a percentage,

acceptable quality level: (AQL) the maximum number of defects per hundred or per cent defective (whichever is specified—it should be noted that these two measures are not the same thing) which will be considered satisfactory by the customer,

sample: a selection of items to undergo test or inspection taken from the lot or batch; there are a number of prescribed ways of selecting the sample number described briefly below,

inspection level: the inspection level relates the sample size to the batch size; as the batch increases the sample size is increased but not necessarily proportionally,

operating characteristic (*o.c.*) *curve:* a curve plotting the percentage of batches expected to be acceptable against the values of per cent defective; this curve is a plot based on probability theory and its shape is determined by batch and sample sizes, *i.e.* the inspection level; the AQL may be marked on the per cent defective axis. This is discussed in a little more detail below.

Basic technique

In the guide given by the Government Inspectorate, a number of sampling plans are presented. These relate batch sizes, sample sizes and the AQL and give the 'accept' or 'reject' figures to be adopted for a particular inspection. To select a sampling plan the following details must be known: the batch size, the AQL, the inspection level, the inspection type and the type of sampling to be used. There are three kinds of inspection, normal, reduced and tightened and, as the names imply, they are a measure of the customer's confidence in the manufacturer. Also, there are four types of sampling: single, double, multiple or sequential, and, again as implied, this describes the time sequence in the selection of sample sizes.

A number of possible sampling plans are presented and the customer and manufacturer negotiate choice and types when the contract is drawn up. In all these negotiations care is taken that on the one hand the customer is not taking a high risk in accepting the goods on the conditions laid down, and on the other that the manufacturer is not taking unnecessary precautions to maintain quality during production. It must be remembered that the final cost of the product must take into consideration the outlay involved in quality control and inspection techniques.

An illustration of compromise may be seen by examination of typical o.c. curves illustrated in Fig. 10.1 which shows curves for three sets of conditions. Figure 10.1(*a*) shows the ideal (and unattainable) curve. It can be seen that the probability of acceptance of items better or equal to the AQL, *i.e.* having a per cent defective figure less than or equal to the AQL, is 100 per cent, whilst the probability of acceptance of items worse than the AQL is zero. Figure 10.1(*b*) shows a very low probability of acceptance of 'worse' items—which is, of course, pleasing to a customer—but an equally low probability of acceptance of items slightly better than the AQL. In other words, the quality must be maintained at a much higher level than that which is acceptable to ensure acceptance. This could mean that the manufacturer may be going to unnecessary expense. Figure 10.1(*c*) shows a curve which favours the manufacturer since there is a high probability of acceptance of 'better' items but the customer incurs a high risk since there is an equally high probability of acceptance of slightly worse items. The o.c. curves result from a statistical analysis based on levels of inspection, batch and sample sizes, sampling methods and so on. The manufacturer and customer thus negotiate a set of inspection and testing conditions which is to their mutual satisfaction.

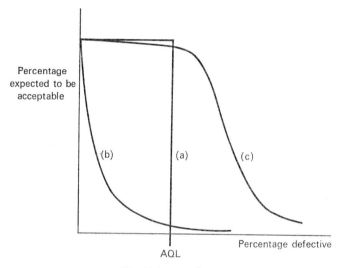

Fig. 10.1   o.c. Curves.

10.5   TESTING TECHNIQUES: RESULTS, TABULATION AND ANALYSIS

As was described in the preceding section, statistics and the mathematical theory of probability play a large part in testing and inspection methods. They are also used to a considerable degree in analysis of test results and prediction of true values. Again the scope of this text precludes a detailed discussion and only fundamentals will be described. The application of statistical mathematics to reliability analysis is described in more detail in texts named in the bibliography at the end of the chapter.

As with all specialist studies the best beginning is an examination of vocabulary and basic methods. In determining the value of any particular variable, for example the scale reading on an instrument undergoing test, it is obviously advisable not to rely on a single observation. Consequently a number of observations of the same value are made and an analysis may then be carried out to determine the true reading and thus, after all other errors have been eliminated, to determine the instrument error if any. For such a set of observations there are a number of standard methods of tabulation and display and certain standard terms describing the set or characteristics of the set.

Presentation of data

If only a small number of observations is made, data presentation may be made by means of a 'dot diagram' in which a dot is marked on a calibrated scale indicating the measured variable. An example of this is shown in Fig. 10.2 for readings on a voltmeter of 4·15 V, 4·17 V, 4·13 V, 4·14 V, 4·15 V, 4·18 V, 4·17 V, 4·16 V and 4·15 V, *i.e.* a set of nine observations in all. As the number of observations increases, however, the usefulness of the dot diagram as a means of

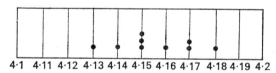

4·1 4·11 4·12 4·13 4·14 4·15 4·16 4·17 4·18 4·19 4·2

Fig. 10.2 Dot diagram.

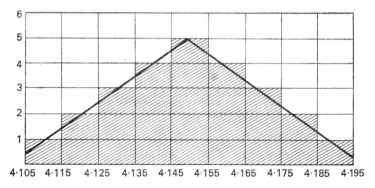

Fig. 10.3 Frequency histogram.

display falls off and the more usual method is then the frequency distribution. In this the range in variation of the observed values is divided into equal intervals or classes and the number of observations lying within each interval (the frequency) is plotted against the range of values of the variable. If this is plotted as a bar diagram as in Fig. 10.3 for the observations described below, it is known as a frequency histogram.

Figure 10.3 shows a frequency histogram for the following observations arranged in numerical sequence:

Class interval	Frequency	Observations
4·105 to 4·115	1	4·11
4·115 to 4·125	2	4·12
		4.12
4·125 to 4·135	3	4·13
		4·13
		4·13
4·135 to 4·145	4	4·14
		4·14
		4·14
		4.14
4.145 to 4.155	5	4·15
		4·15
		4·15
		4·15
		4·15
4·155 to 4·165	4	4·16
		4·16
		4·16
		4·16
4·165 to 4·175	3	4·17
		4·17
		4·17
4·175 to 4·185	2	4·18
		4·18
4·185 to 4·195	1	4·19
total 25		sum 103·75

Standard terms

The *arithmetic mean* (or mean) is the sum of the observations divided by their number. For the example given above this is 103·75 divided by 25, *i.e.* 4·15. It can be shown that the mean value is the closest approximation to the true value, the accuracy improving as the number of observations is increased.

The *mode* is the most frequently occurring value. Again, for the example given it is 4·15. For a symmetrical set of observations such as the one given, *i.e.* with 'low' values and 'high' values being symmetrically displaced about the mean, the mean and mode have the same value.

Accuracy of measurement is a measure of how the values lie with respect to the true value, whereas *precision* in measurement is a measure of the closeness of a group of values. A closely grouped set of observations lying near to the true value would indicate both accurate and precise measurement, if such a group was away from the true value the measurement would be precise but inaccurate. An indication of precision is the way in which observations in a set are distributed about the mean; such an indication is called *scatter* or *dispersion*. The overall range of a set of values (difference between largest and smallest) is one way of describing scatter.

Deviation is the difference between any observation and the mean. If all deviations are added, irrespective of whether they are low or high, and divided by the number of observations the result is the *mean deviation*. The mean deviation indicates the average error (between observed values and mean value) but does not, however, adequately show how the errors are distributed, *i.e.* the scatter of the observations. A more useful indication is the r.m.s. or *standard deviation* which is the square root of the mean of the squares of the deviations. The parameter looks rather involved but is, however, one of the most useful of all those associated with statistical analysis. The symbol for standard deviation is σ_s (sigma).

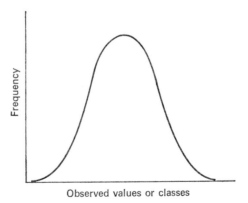

Observed values or classes

Fig. 10.4 Normal distribution curve.

The normal distribution

If a very large number of observations is made which vary symmetrically about the mean value and a frequency histogram (Fig. 10.3) is plotted with very small class intervals, the resultant shape is as shown in Fig. 10.4. This bell-shaped curve is called a *normal* or *Gaussian* distribution, and it is found that a very great number of kinds of observations will produce a distribution approximating to the normal. In fact measures of departure from normalcy often act as a useful guide to the presence of errors other than random as explained below.

It is found that for a normal curve there is a definite relationship between the distribution of observed values about the mean and the standard deviation as defined above. The relationship is that if the mean value is denoted by M and the standard deviation by σ_s then

68·27 per cent of the observations lie between $M - \sigma_s$ and $M + \sigma_s$
95·45 per cent lie between $M - 2\sigma_s$ and $M + 2\sigma_s$ and
99·73 per cent lie between $M - 3\sigma_s$ and $M + 3\sigma_s$.

Thus it can be said that the probability of any observation lying between $M \pm \sigma_s$ is 0·6827, between $M \pm 2\sigma_s$ is 0·9545 and between $M \pm 3\sigma_s$ is 0·9973 and the normal curve is often referred to as a *normal probability* curve. (It should be noted that probability is not certainty. An illustration of this is in tossing a coin where the probability of 'heads' is 0·5; the larger the number of throws the more the observed results correspond to those predicted by probability.)

The normal curve is well known and has been analysed in great depth. Consequently if tests show that a distribution approaches the normal a number of assumptions may then be made. A particularly useful application of this is in the estimation of random error, since, from above, for any one observation in a set of normally distributed observations there is a

0·6827 probability that the error lies between $\pm \sigma_s$
0·9545 probability that the error lies between $\pm 2\sigma_s$
0·9973 probability that the error lies between $\pm 3\sigma_s$

Further, the closer that the results approximate to normalcy, the more we can be sure that systematic error (as described later) has been eliminated.

Example 10.1

Calculate the arithmetic mean and standard deviation of the following set of observations. Assuming normal distribution, between what values is there a 0·95 probability that any further observation would lie?

Frequency	Observation value
2	44
3	46
4	48
5	50
4	52
3	54
2	56

The observations and required calculations are tabulated below:

Frequency	Value	Deviation
2	44	6
3	46	4
4	48	2
5	50	0
4	52	2
3	54	4
2	56	6
Total 23	1150	

Sum of observations:

$(2 \times 44) + (3 \times 46) + (4 \times 48) + (5 \times 50) + (4 \times 52)$
$$+ (3 \times 54) + (2 \times 56)$$
$$= 88 + 138 + 192 + 250 + 208 + 162 + 112 = 1150$$

Number of observations is 23. Thus mean value is

$$\frac{1150}{23}$$

i.e. 50. The sum of the squares of the deviations is

$(2 \times 6^2) + (3 \times 4^2) + (4 \times 2^2) + (4 \times 2^2) + (3 \times 4^2) + (2 \times 6^2)$
$$= 72 + 48 + 16 + 16 + 48 + 72 = 272$$

The standard deviation is thus

$$\left(\frac{272}{23}\right)^{\frac{1}{2}}$$

i.e. 3·44, and there is a 0·95 probability that any further observation would lie between

$$50 \pm 2 \times 3·44, \qquad i.e. \text{ between } 43·12 \text{ and } 56·88$$

Notice in this example the use of the frequencies of observed values and that the calculation actually gives a 0·9545 probability level rather than 0·95.

10.6   TESTING TECHNIQUES: TYPES OF ERROR

The errors occurring in any observation or series of observations may be categorised into two types: *random* and *systematic*.

Random errors, as the name implies, are unpredictable as to size or nature ('low' or 'high'). They occur completely at random and as such for any set of observations of the same nominal value tend to be self-compensating, *i.e.* the probability is that 'high' errors compensate for 'low'. This is one of the reasons why the arithmetic mean of a set of observations is the best value (in the sense that it is closest to the true value), since the compensating ability has full opportunity to work in this case.

Systematic errors are predictable errors and their cause may be determined and eliminated. They occur regularly, *i.e.* with each observation, and as such tend to be cumulative. They may be sub-categorised into observer error, instrument error and natural error. The names are fairly self-explanatory in that observer error is due to some characteristic of the person making the observations, for example a tendency to read 'high' or 'low' when estimating the position of a pointer between calibrations. Observer error is usually compensated by a change of observer and making several sets of observations. Instrument error is due to some defect within the instrument used for measuring. In the case of instruments undergoing tests it is this error which is being sought. A detailed discussion of instrument error for the more common instruments used under various conditions is contained in Chapter 8. Natural error is due to natural or physical phenomena such as expansion of materials when heated, the influence of atmospheric pressure or content, etc. Once found and distinguished by cause, compensation may be made for this type of error.

Presence of systematic errors is indicated by the 'normal' curve being shifted either to left or right in a *non-uniform* manner, *i.e.* all points unequally. If a preliminary test run indicates departure from normalcy, a search for such an error cause may immediately be made.

10.7 ESTIMATION OF EXPERIMENTAL ERROR

As was stated above, for a normal distribution curve of repeated readings, the skewness of the curve, *i.e.* departure from normalcy, gives an indication of systematic error. Examination of the curve once normalised (*i.e.* once systematic errors have been eliminated as far as possible) using statistical methods then gives an estimate of random errors within the observations.

The final error in results obtained from test procedures involving more than one instrument or component is obviously dependent on the individual errors of each instrument or component. It can be shown* that, if the final result depends upon the sum, product or quotient of two or more factors, then the final relative error is equal to the sum of the individual relative errors, where relative error means the actual error expressed as a fraction or percentage of the whole. Thus in the measurement of resistance by the voltmeter–ammeter method, for example, the result depends upon the division of the voltmeter reading by the ammeter reading, *i.e.* the quotient of two factors. The resultant relative error for the resistance is then the sum of the relative error of the voltmeter and that of the ammeter. Suppose, for example, that each instrument has an inherent error of ± 1 per cent, the resistance result could then have an error of ± 2 per cent. These errors are systematic since they are due to the instrument construction, etc. Additional errors which may or may not occur with each observation would be classified random, as explained in Section 10.6.

Random errors are analysed using statistical methods. As was stated earlier the best value of any result obtained from a series of observations is the mean value. The anticipated error in any observation then has a 68·27 per cent probability of occurring within the limits $\pm \sigma_s$ (where σ_s is the standard deviation), a 95·45 per cent probability of occurring within the limits $\pm 2\sigma_s$, and so on. The word, 'error' in this context means a deviation between the observed value and the mean value which is assumed true. Now this assumption is only valid if all errors are truly random and they are evenly distributed. It has been shown that this can only be assumed for an infinite number of observations. In other cases, *i.e.* in practice, the mean value may, in fact, not be the true value. One way to obtain a more accurate estimate of the true value is to repeat *sets* of observations, determine the mean of each set and plot these mean values. The *mean* of the means then is even closer to the true value than a single mean.

---

* For example in *Principles of Electrical Measurements* by Buckingham and Price, English Universities Press.

The statistical methods applied to a single set of observations can then be applied to the set of mean values and a *mean square error* (corresponding to the standard deviation of a set of observations) can be obtained. The mean square error of the mean is not called standard deviation but is given a special name *standard error*, symbol σ_m. The standard error is a measure of how far from the true value (determined from several *sets* of observations) the mean value (determined from any *single set* of observations) lies. The probability percentages still apply to the standard error since it can be shown that for a number of sets of normally distributed observations the mean values thus obtained will also be distributed normally. In this context we refer to being *confident* that values will lie within a certain range, the degree of confidence being determined by the probability. Thus we can be

68·27 per cent confident that any mean value obtained will lie within $\pm\sigma_m$ from the true value
95·45 per cent confident that it will lie within $\pm2\sigma_m$ from the true value, and so on.

Fortunately it has been shown in statistical works that it is not necessary to repeat sets of observations since the standard error (mean square error of the mean) is related to the standard deviation (mean square error of a single observation) by the equation

$$\sigma_m = \frac{\sigma_s}{(n)^{\frac{1}{2}}} \tag{10.1}$$

where n is the number of observations in a single set.

Example 10.2
Determine the levels between which we can be 68·27 per cent confident that the true mean value of Example 10.1 will lie.

From Example 10.1

mean value is 50
standard deviation, σ_s, is 3·44
number of observations, n, is 23

Thus from eqn. (10.1) standard error σ_m is given by

$$= \frac{\sigma_s}{(n)^{\frac{1}{2}}}$$
$$= \frac{3\cdot44}{(23)^{\frac{1}{2}}}$$
$$\simeq 0\cdot717$$

Thus, since we can be 68·27 per cent confident that the true value lies between $\pm\sigma_m$ of the mean value obtained, we can be 68·27 per cent confident that the true value lies between $50 - 0\cdot717$ and $50 + 0\cdot717$, *i.e.* 49·283 and 50·717.

Occasionally the term 'probable error' is used. This is an error equal to $0\cdot6745\sigma_s$ if error means deviation between a single observation and the mean, or $0\cdot6745\sigma_m$ if 'error' means the deviation between the mean and the true value. There is a 0·5 probability that the error lies within $\pm0\cdot6745\sigma_s$ (or $\pm0\cdot6745\sigma_m$) or in other words there is an equal probability of the error lying within this range as outside it. In the above example the probable error of any single observation is $0\cdot6745\sigma_s$, *i.e.* $0\cdot6745 \times 3\cdot44$ which equals 2·32, and there is a 50 per cent probability that the deviation between any single observation and the mean will lie between $\pm2\cdot32$, so that equally any single observation will lie between $(50 - 2\cdot32)$ and $(50 + 2\cdot32)$. The probable error of the mean is $0\cdot6745\sigma_m$, *i.e.* $0\cdot6745 \times 0\cdot717$ which equals 0·484, and there is a 50 per cent probability that the deviation between the mean and the true value will lie between $\pm0\cdot484$, so that equally the true value will lie between $(50 - 0\cdot484)$ and $(50 + 0\cdot484)$.

Example 10.3
The power in a resistive load is measured simultaneously by the voltmeter–ammeter method and by a wattmeter. The voltmeter is a 0–30 V instrument having a linear scale divided into 120 divisions, the ammeter is a 0–20 A instrument with a scale divided into 200 divisions and the wattmeter is a 0–300 W instrument with a scale divided into 300 divisions. All three instruments may be read to 0·5 of a division and may be assumed to have an inherent (instrument) error of ±1 per cent. Determine the limits of error of the two methods when the readings are voltmeter 9·95 V, ammeter 10 A and wattmeter 99·85 W.

Possible reading error of voltmeter is 0·5 of a division and since one division corresponds to 0·25 V is thus $0\cdot5 \times 0\cdot25$, *i.e.* 0·125 V. The relative error of the voltmeter is thus

$$\frac{0\cdot125}{9\cdot95} \times 100 \text{ per cent}$$

i.e. 1·25 per cent. The possible reading error of the ammeter is 0·5 of a division, *i.e.* $0\cdot5 \times 0\cdot1$ A, *i.e.* 0·05 A.

The relative error of the current reading is

$$\frac{0\cdot05}{10} \times 100 \text{ per cent}$$

i.e. 0·5 per cent. The power value is determined by the product of these readings and thus is subject to a relative error equal to the sum of the individual relative errors. Thus relative error of power value is 0·5 + 1·25, *i.e.* 1·75 per cent. In addition there is an inherent error of ±1 per cent in each instrument so that total relative error possible is 1 + 1 + 1·75 *i.e.* 3·75 per cent.

For the wattmeter, the reading error could be 0·5 of a division, *i.e.* 0·5 × 1 W. The relative error is thus

$$\frac{0·5}{99·85} \times 100$$

i.e. 0·5008 per cent. The inherent error could be 1 per cent. The total error possible is thus (1 + 0·5008) per cent, *i.e.* 1·5008 per cent. The wattmeter method is preferable in this case.

Example 10.4

The following set of observations was made of the reading on an ammeter. Determine the probable error in any one of these observations and the limits between which the true value lies at a 0·6827 level of probability.

4·9, 4·9, 4·95, 4·96, 4·99, 5, 5, 5, 5·01, 5·02, 5·02, 5·05, 5·1

To obtain a solution it is necessary to find the standard deviation which in turn necessitates finding the mean value.

The solution is laid out as follows:

	Reading	Deviation from mean	Square of deviations
	4·90	0·09	0·0081
	4·90	0·09	0·0081
	4·95	0·04	0·0016
	4·96	0·03	0·0009
	4·99	0·0	0·0
	5·00	0·01	0·0001
	5·00	0·01	0·0001
	5·00	0·01	0·0001
	5·01	0·02	0·0004
	5·02	0·03	0·0009
	5·02	0·03	0·0009
	5·05	0·06	0·0036
	5·10	0·11	0·0121
sum	64·90		0·0369

Mean value

$$= \frac{64 \cdot 9}{13}$$

$$= 4 \cdot 99$$

Standard deviation

$$= \left(\frac{0 \cdot 0369}{13} \right)^{\frac{1}{2}}$$

$$= 0 \cdot 053\ 27$$

The probable error of a single observation

$$= 0 \cdot 6745 \times 0 \cdot 053\ 27$$

$$= 0 \cdot 035\ 93$$

The standard error

$$= \frac{\text{standard deviation}}{(\text{number of observations})^{\frac{1}{2}}}$$

$$= \frac{0 \cdot 053\ 27}{(13)^{\frac{1}{2}}}$$

$$= 0 \cdot 014\ 78$$

There is a 0·6827 probability that the true value lies between (mean $\pm$ standard error), *i.e.* in this case between (4·99 + 0·014 78) and (4·99 − 0·014 78). The required limits are thus 5·004 78 and 4·975 22 (probably 5·00 and 4·97 to the reading limits imposed by the instrument).

10.8   TESTING OF INSTRUMENTS

The various types of instrument, methods of use and common errors were considered in some detail in Chapter 8. It is recommended that this chapter be read in conjunction with this section.

Standards of performance under test of electrical instruments in general are laid down in British Standard publication BS 89:1954. In this all instruments are divided into two categories: *precision* and *industrial*. The individual specifications are detailed and the reader is referred to BS 89 for complete definition of any point. Generally, however, precision instruments are those which are intended for testing where high accuracy is required. They are fitted with knife edge pointers, finely divided scales with mirrors (to avoid parallax error: *see* Chapter 8) and are used horizontally. The specified error limits are small. Industrial instruments are intended for use in workshop and factory testing *in situ*, the scales and pointers being determined by the particular use. They may be used horizontally or vertically depending on type and their error limits vary, increasing as the required accuracy is reduced. In general they are somewhat more robust and are invariably calibrated (or checked once calibrated) using a precision instrument of the same type.

In the calibration of an industrial grade instrument using a precision grade instrument, the recommended method of checking scale reading is to adjust the industrial instrument reading to a particular level and determine the error by reading the precision instrument. One reason for this is that it is easier to obtain an accurate reading of a pointer off a main calibration mark on a precision instrument because of its finely divided scale. The reverse procedure, *i.e.* setting the precision instrument and reading the industrial instrument is often more convenient but it should be remembered that accuracy may thus be impaired. In calibration or comparison of instruments it is often useful to draw a calibration curve. In this the instrument reading as observed is plotted against the true value determined either by a precision instrument or even more accurate measurement method (*see* Chapter 9), and a series of points is obtained. The graph is obtained by joining point to point, and a smooth curve should *not* be drawn through the points as is often done in experiments involving continuous variables. The usefulness of the calibration curve is considerably improved by taking a large number of points for comparison. The correct value of the variable when using the calibrated instrument may then be read off the calibration curve, interpolating between points if necessary. If error is plotted against true value the resultant graph is the instrument error curve and the same procedure applies.

Example 10.5
An industrial voltmeter and a precision grade voltmeter gave the following readings during a calibration test. Draw up an error table and plot the correction curve. What correction should be applied to the industrial instrument when it reads 55 V?

Industrial	0	10	20	30	40	50
Precision	0	10·5	20·5	31	39·5	49·5
Error	0	−0·5	−0·5	−1	+0·5	+0·5

Industrial		60	70	80	90	100
Precision		61	70·5	80	89·5	99·5
Error		−1	−0·5	0	+0·5	+0·5

The correction curve (error graph) is shown in Fig. 10.5. By interpolation at 55 V reading the industrial grade is reading 0·25 V low.

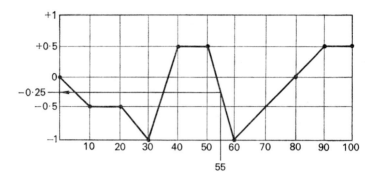

Fig. 10.5   Calibration curve for Example 10.5.

10.9   TESTING OF COMPONENTS

Methods of determination of resistance, inductance and capacitance were described in detail in the previous chapter. For accurate results and to determine error the methods outlined in Sections 10.5, 10.6 and 10.7 are employed.

10.10   TESTING OF SMALL MACHINES

Machine characteristics of interest are torque/speed, torque/load and speed/load relationships, efficiency, safety of operation and reliability. Testing must thus cover behaviour on and off load, during starting and after running for some time. The parameters which require to be measured include voltage, current, power, frequency, speed, torque, temperature and resistance. A number of these parameters were discussed earlier and a considerable part of machine testing involves the applications of methods discussed in Chapter 9 and instruments discussed in Chapter 8.

For the measurement of voltage, current and power, high grade laboratory instruments are used. These are the 'hybrid' variety discussed in Chapter 8 having combined characteristics of the two types 'precision' and 'industrial'. Frequency measurement may be made using one of the methods described in Chapter 9 or for certain applications may be safely assumed constant at 50 Hz. A useful method of checking on this is to note the accuracy of electric clocks in the vicinity of the test area. (Clock drives employ synchronous motors whose speed is determined by frequency.)

Methods of measuring speed, torque and temperature are discussed below.

Speed

Speed is normally measured by one of three possible methods: the revolution counter, the tachometer or the stroboscope. The revolution counter is a relatively crude instrument and requires the additional use of a stopwatch. The counter takes the form of a rubber tipped shaft driving a disc through appropriate gears. The shaft is inserted into the indentation at the end of the motor shaft and the disc rotates at a reduced speed which is known. If a stopwatch is started on insertion the number of revolutions per unit time may thus be determined. The method is rather crude and liable to error since it involves simultaneous starting of two instruments. The tachometer includes both the stopwatch mechanism and counter in the same instrument, both sections being started by the one button. After a predetermined period of time a clutch mechanism disengages the drive mechanism and the scale is appropriately calibrated so that the pointer automatically reads speed. The stroboscope employs a method which is now used in a variety of other ways, including observation of work being carried out on moving objects, for example in a lathe. The method is to illuminate the moving shaft for very short periods. If the same part of the shaft is illuminated every time the light flashes the shaft appears to stand still and its speed may be determined from the frequency of the flashes. (It is this effect which has to be borne in mind when fluorescent lighting is used in workshops.) The stroboscope contains an oscillator coupled to a neon lamp, the oscillator frequency being adjustable via a calibrated scale.

When alternating current is applied to a neon lamp it flashes twice per cycle, once each time the wave reaches a peak. The number of flashes per unit time is thus related to the applied frequency. When using the stroboscope the frequency is adjusted until the shaft appears to stop and the speed may then be read off the scale directly. The method has a high accuracy of measurement.

Torque

Torque of motors is usually measured by the device which is used to apply the load to the machine. The two methods commonly used employ mechanical loading (via friction) or electrical (via a generator being driven by the motor under test). Friction loading systems such as the Prony brake described below have the disadvantage that excessive heat may be generated at the load end of the machine under test. With electrical loading the power generated due to loading may be dissipated as heat in resistive loads placed some distance away.

The Prony brake consists of an arm which is clamped via a friction lining (of the brake shoe variety) to a pulley fixed to the machine shaft. As the shaft rotates the pulley rubs against the friction lining and the arm will tend to follow the shaft rotation. If sufficient torque is applied to the arm by attaching weights to one end the arm will just stop short of rotating, *i.e.* will lift the weights. When the brake arm is horizontal the torque exerted by the machine is equal and opposite to the torque exerted by the brake arm. This is determined directly by the product of weight × distance along the arm between machine centre and point of attachment of the weights. The heat produced by friction is considerable and arrangements to cool the brake lining may be necessary.

A generator specially made for torque measurement and load tests is called a dynamometer. In the conventional generator the torque applied to the armature makes the rotor turn and there is an equal torque applied to the stator. Since the stator is fixed this is of no importance. In the dynamometer the stator is free to move through a small angle and when the armature rotates the stator will tend to follow it. An arm is attached to the stator and, as with the Prony brake, weights may be suspended from the arm to restrain movement. As before when the arm is horizontal the torques are equal and opposite and the machine torque may be calculated. On certain more expensive machines the torque arm is attached via a spring to a calibrated scale fixed to the ground and the torque may be read off directly when the spring tension, which is adjustable, is held at the correct level for the torque arm to be held horizontal. The dynamometer output is fed to a resistor bank and the heat generated may be safely dissipated.

Temperature

The permitted temperature rise of any machine during a run is specified by the appropriate British Standard (*e.g.* BS 170 for fractional horsepower machines), and it is necessary to test this aspect to maintain performance to within specification. Temperature may be measured either by the usual glass thermometer containing an appropriate temperature sensitive fluid or by a thermocouple. The thermocouple is a junction of dissimilar metals which when heated causes a voltage to be set up between the two wires. The size of this voltage is related to the temperature of the junction, and instruments are available having a calibrated scale and test probe containing the junction. In some respects this method is better than that using the thermometer in that it is easier to locate the junction within the machine. Temperature is taken at the various parts of the machine as stipulated by the appropriate British (or American) standard. An alternative method of measurement of the temperature rise of the armature conductors is to measure resistance, using one of the methods described in Chapter 9, before and after a run, and use the standard equation relating resistance to temperature for a specified material.

10.11   USE OF LOGARITHMIC UNITS IN TESTING

A considerable number of tests carried out on instruments, components and machines involve the comparison of two quantities or two magnitudes of the same quantity under different conditions. The comparison thus involves the use of a ratio. It is often convenient to express such a ratio using logarithms.

The most commonly used logarithmic ratio is the *bel* and its sub-unit the *decibel* which should be strictly applied to power levels (or voltage or current levels with the same load), but is often erroneously applied to other quantities. The bel and decibel are defined precisely by British Standard 204 to which the reader is referred for detailed discussion.

A ratio of power levels expressed in bels is the logarithm to the base of 10 of the ratio. Thus for two power levels one of which is twice the other, for example, the ratio is log 2 bels, *i.e.* 0·3010 bels abbreviated 0·301 B. Alternatively, the ratio may be expressed as 0·5 since one level is half the other. In bels this is log 0·5 which, by definition, is $-1 + 0·6990$ or $-0·3010$ B. The same figure is obtained regardless of which level is taken as numerator but the sign changes. Thus it is said that the larger level in this case is 0·3010 B *up* on the smaller level or, alternatively, the smaller level is 0·3010 B *down* on the larger level.

A more convenient unit of ratio is the decibel, abbreviated dB or db, which is one-tenth of a bel. The ratio of two power levels expressed in decibels is thus 10 log (ratio) dB.

Example 10.5
Express the following power level ratios in decibels.
(*a*) 2, (*b*) 0·5, (*c*) 3, (*d*) 0·3333, (*e*) 5, (*f*) 0·2.

(*a*)   10 log 2 = 3·01 dB or 3·01 dB up.

(b) $10 \log 0.5 = \bar{1}.6990$
$$= -1 + 0.699$$
$$= -3.01 \text{ dB or } 3.01 \text{ dB down.}$$

(c) $10 \log 3 = 4.771 \text{ dB up.}$

(d) $10 \log 0.3333 = 10(\bar{1}.5228)$
$$= 10 \times (-1 + 0.5229)$$
$$= 4.77 \text{ dB down.}$$

(e) $10 \log 5 = 6.99 \text{ dB up.}$

(f) $10 \log 0.2 = 10(\bar{1}.3010)$
$$= 10 \times (-1 + 0.3010)$$
$$= 6.99 \text{ down.}$$

If two power levels P_1 and P_2 are compared, where

$$P_1 = I_1{}^2 R_1$$

or

$$= \frac{V_1{}^2}{R_1}$$

and

$$P_2 = I_2{}^2 R_2$$

or

$$= \frac{V_2{}^2}{R_2}$$

then the ratio of P_1 to P_2 expressed in decibels is

$$10 \log \frac{P_1}{P_2} \qquad (10.2)$$

which, from above, equals

$$10 \log \frac{I_1{}^2 R_1}{I_2{}^2 R_2}$$

$$= 10 \log \left(\frac{I_1}{I_2}\right)^2 \frac{R_1}{R_2}$$

$$= 20 \log \frac{I_1}{I_2} \quad \text{if} \quad R_1 = R_2 \qquad (10.3)$$

or alternatively

$$10 \log \frac{P_1}{P_2} = 10 \log \frac{V_1{}^2}{R_1} \frac{R_2}{V_2{}^2}$$

$$= 10 \log \left(\frac{V_1}{V_2}\right)^2 \frac{R_2}{R_1}$$

$$= 20 \log \frac{V_1}{V_2} \quad \text{if} \quad R_2 = R_1 \qquad (10.4)$$

Example 10.6
A signal generator is feeding a 1 mV signal into a 600 Ω load. Calculate the new input voltage if the generator attenuator is switched to −20 dB.

Since the load remains the same, eqn. (10.4) m ay be used to give

$$-20 = 20 \log \frac{V_x}{10^{-3}}$$

where V_x is the new voltage input, *i.e.*

$$\log \frac{V_x}{10^{-3}} = -1$$

thus

$$\log \frac{V_x}{10^{-3}} = \bar{1} \cdot 0000$$

and

$$\frac{V_x}{10^{-3}} = 0 \cdot 1$$

and

$$V_x = 0 \cdot 1 \times 10^{-3} \text{ volts}$$

$$= 0 \cdot 1 \text{ mV}$$

A 20 dB reduction is thus equal to a 99/100 or 99 per cent reduction in output.   9/10   or   90%.

Example 10.7
Calculate the voltage or current ratios corresponding to (a) 6 dB up, (b) 3 dB down.

The rates will be the same for voltage or current since eqns. (10.3) and (10.4) have the same form. Thus

(*a*) $$6 = 20 \log (\text{ratio})$$

and

$$\text{ratio} = \text{antilog} \frac{6}{20}$$

$$= \text{antilog } 0\cdot3$$

$$= 2 \text{ (approximately)}$$

Note that a 6 dB change in voltage or current corresponds to a 3 dB change in power.

(*b*) $$-3 = 20 \log (\text{ratio})$$

and

$$\text{ratio} = \text{antilog} \left(-\frac{3}{20} \right)$$

$$= \text{antilog } (-0\cdot15)$$

$$= \text{antilog } \bar{1}\cdot85$$

$$= 0\cdot7079$$

i.e. the voltage or current is reduced to $0\cdot7079$ of the original for a 3 dB down adjustment.

Notice that the wholly negative number $-0\cdot15$ must be changed to the usual logarithmic form (*i.e.* a negative characteristic and positive mantissa) before tables can be used. Occasionally one finds the term decibel applied to voltage or current levels where the load value changes. The correct unit recommended by BSI for these instances is *decilog* (*see* BS 204). It should be noted that attenuators or other controls calibrated in decibels are only accurate for voltage or current if the load remains constant.

Use of logarithmic graph paper

Occasionally, in the plotting of characteristics it is found that a variable increases its value from zero over a wide range. One example is an amplifier frequency response (*see* Chapter 7) where the frequency range may extend from 50 Hz to 20 kHz and one is equally interested in the range 50 Hz to, say, 3 kHz as in the 15 kHz to 20 kHz region. In such cases a linear scale may be too large and inconvenient. A logarithmic scale, in which equal increments along the axis indicate the variable raised to the power ten, may then be used. An example of such a scale follows:

Scale length (cm)	10	20	30	40	50	60
Frequency	10 Hz	100 Hz	1 kHz	10 kHz	100 kHz	1 MHz

Care must be taken when plotting graphs using such scales and their nonlinear nature must be borne in mind.

FURTHER READING

As was stated at the beginning of this chapter, the subject of testing methods is so wide that readers may care to continue their studies with further reading. A selected list of texts follows:

(1) *Quality Control Handbook*, J. M. Juran (Ed); McGraw-Hill.
(2) British Standard 89:1954 *Electrical Indicating Instruments*.
(3) *Engineering Measurements*, by B. Austin Barry; John Wiley & Sons.
(4) *Introduction to Reliability Engineering*, by Rhys Lewis; McGraw-Hill.
(5) *Fractional Horsepower Motors*, by Stuart F. Philpott; Chapman & Hall.

PROBLEMS ON CHAPTER TEN

The problems in this exercise cover the theory contained in Chapters 8, 9 and 10.

(1) Describe a test made to check the accuracy of a moving coil workshop-type voltmeter. During such a test the following results were obtained:

Workshop voltmeter reading	0	25	50	75	100	125	150
Superior grade voltmeter reading	0	24	49·5	76	101	124	151

Prepare a table of errors and a correction curve. Ignore any errors in the superior grade instrument. What correction must be applied to the workshop instrument when it reads 85 V?

(2) List the advantages and disadvantages of moving coil, moving iron and electrodynamic indicating instruments. State which could be used to best advantage for the following measurements. Explain the choice.

(a) a dc resistive circuit,
(b) an ac resistive circuit supplied at:

(i) 10 Hz, (ii) 100 Hz, (iii) 1000 Hz, (iv) 1 MHz.

(3) Describe the voltmeter–ammeter method of measuring (a) high resistance, (b) low resistance. In such a test the voltmeter resistance was 25 000 Ω, the ammeter resistance 0·4 Ω. The measured resistance was 20 Ω. Estimate the percentage error of measurement if method (a) is used. Ignore reading errors.

(4) Definite and discuss the meaning of the following terms:

(a) specification,   (b) acceptable quality level,
(c) inspection,   (d) sampling.

(5) What is meant by probable error?
Twenty-eight measurements of a certain variable were distributed in the following manner:

No. of measurements	Value of measurements
1	1400
4	800
7	900
8	1000
5	1100
3	1200

Determine the mean value of this measurement and estimate the probable error of the mean.

(6) Ten readings of a voltmeter were taken as follows:

9·8 V, 9·8 V, 9·9 V, 9·9 V, 10 V, 10 V, 10 V, 10·1 V, 10·2 V, 10·2 V.

Calculate:

(a) the mean value,
(b) the standard deviation,
(c) the limits between which there is a 68 per cent probability that any new reading would lie.

(7) Discuss random and systematic errors in measurement. The value of a resistance was measured by connecting a 20 V voltmeter across the resistor and a 5 A ammeter in series with the combination. The voltmeter and ammeter readings were respectively 10 V and 2 A. The scale of each instrument had 200 divisions, and could be read with certainty to ±0·5 division. The instrument error of each instrument lay between ±1 per cent. Calculate the value of the resistance and the possible error limits. The voltmeter current may be ignored.

(8) Describe the two methods of connection of a wattmeter in circuit. A 0–150 W wattmeter having a 5 Ω current coil and a 10 000 Ω voltage coil is used to measure the power taken by a 100 Ω load connected across a 100 V supply. Estimate the relative error of each method of wattmeter connection.

(9) Define the decibel. A signal generator provides a 100 mV signal into a 600 Ω load with the attenuator set at 0 dB. Calculate:

(a) the signal voltage,
(b) the load current,
(c) the load power when the attenuator is set at:

    (i) 20 dB down, (ii) 40 dB down.

(10) Describe methods of measuring speed, torque and input power for a performance test on a fractional horsepower motor.

(11) The power absorbed by a series tuned circuit at resonance is 0·1 W. The resonant frequency is 100 kHz, and the circuit Q factor is 200. Determine:

(a) the power absorbed at the −3 dB points on the resonance curve,
(b) the bandwidth,
(c) the frequencies at which the −3 dB points occur.

Test Papers

TEST PAPER ONE

(1) Explain with the aid of neat sketches how a shunt connected dc generator builds up its open-circuit voltage. If such a generator has lost its residual magnetism how can it be restored?

A separately excited generator on test at 800 rev/min gives the following values:

Induced e.m.f. (V)	12	44	73	98	113	122	127
Field current (A)	0	0·2	0·4	0·6	0·8	1·0	1·2

The machine is then shunt connected and run as a shunt-generator at a speed of 1200 rev/min with a field circuit resistance of 200 Ω. Determine graphically the no-load terminal voltage.

(2) Explain from first principles how a field regulator controls the speed of a dc shunt-connected motor.

A shunt motor runs at 750 rev/min and takes 50 A from a 400 V supply. Determine the speed at which it will run when taking 30 A from a 250 V supply. The resistance of the armature circuit is 0·5 Ω, and the field flux may be assumed to be proportional to the supply voltage.

(3) Explain, with the aid of diagrams, Thévenin's theorem and use it to solve the following problem.

Two batteries are connected in parallel; the e.m.f. and resistance of one battery are 10 V and 1 Ω respectively, and the corresponding values for the other are 11 V and 2 Ω. A resistor of 8 Ω is connected across the battery terminals. Determine the current through this resistor.

(4) Explain, with the aid of sketches, what is meant by resonance in an electrical circuit, and state the essential differences between resonance in a series circuit and resonance in a parallel circuit.

A coil having a resistance of 3000 Ω and a self-inductance of 20 H is connected in series with a capacitor to a 240 V, 50 Hz (c/s) sinusoidal supply. Determine at resonance:

(a) the current flowing in the circuit,
(b) the p.d. across the capacitor,
(c) the Q-factor of the circuit.

(5) Define power factor and explain why capacitors are often used when the power factor is to be improved. A single-phase load when connected to a 240 V, 50 Hz (c/s) supply takes 30 kVA at a lagging power factor of 0·8. Determine the kvar rating of a capacitor which connected in parallel with this load will improve its overall power factor to 0·95 lagging.

Draw a diagram to scale in support of your calculations.

(6) Explain what is meant by the time constant of a circuit in which a resistor is connected in series with an inductor.

A coil has a resistance of 200 Ω and a self-inductance of 10 H and is connected to a dc supply of 50 V of negligible source resistance. Determine the time constant of the circuit, also draw a current/time graph and determine therefrom the coil current 0·04 s from switch-on.

(7) Derive with the aid of a voltage phasor diagram the numerical relationship between line and phase voltages in a star-connected system. Using this relationship prove that three times the power per phase equals $(3)^{\frac{1}{2}} V_L I_L \cos \phi$ in the case of a star-connected balanced load.

Three coils when connected in star dissipate 2·4 kW at a power factor of 0·6 lagging, when connected to a 416 V, 50 Hz (c/s) 3-phase supply. Determine:

(a) the power dissipated in each coil,
(b) the current in each coil,
(c) the resistance of each coil.

(8) Explain why the current in the primary winding of a transformer varies when the secondary load impedance is altered.

Derive an equation giving the relationship between primary and secondary voltages, turns and currents, assuming the transformer has no losses.

A single-phase transformer has a nett cross-sectional area of 200 cm^2 and the primary winding consists of 1500 turns. The flux density in the core has a maximum value of 1·3 Wb/m^2 and the supply frequency is 50 Hz (c/s). Calculate the r.m.s. value of the induced e.m.f. in the primary winding.

(9) Explain why a dc shunt motor is fitted with

(a) a starter,
(b) a field regulator.

A 220 V dc shunt motor has an armature resistance of 0·25 Ω. Calculate:

(i) the resistance to be connected in series with the armature to limit the armature current to 80 A at starting,

(ii) the value of the generated e.m.f. when the armature current has fallen to 55 A with this value of resistance still in circuit.
Brush voltage drop may be neglected.

This paper included by courtesy of the East Midland Educational Union.

TEST PAPER TWO

(1) Derive with the aid of a phasor diagram the relationship between the line and phase voltages in a star-connected system. Three inductors each of resistance 6 Ω and reactance 8 Ω are connected in star to a 415 V three-phase supply.
Calculate:

(a) the phase impedance,
(b) the phase voltage,
(c) the phase current,
(d) the power taken from the supply,
(e) the power factor of the load.

(2) An inductor having a resistance of 18 Ω and a reactance of 24 Ω is connected in parallel with a capacitor, and these two components are energised from a 250 V single-phase supply of 50 Hz (c/s).
Calculate:

(a) the current taken by the inductor,
(b) the capacitance of the capacitor if the overall power factor is 0·8 lagging,
(c) the current in the capacitor circuit.

Draw a phasor diagram giving the relationship between the applied voltage and the currents in the parallel circuits.

(3) Explain why the secondary voltage of a transformer is (approximately) proportional to its secondary turns.
With the aid of a sketch give constructional details of an oil immersed transformer, and explain:

(a) why oil is used,
(b) why the core is laminated.

(4) With the aid of a neat sketch describe the construction and working principles of a moving coil instrument.
State why such an instrument will give no reading when alternating current at 50 Hz (c/s) is passed through the coil.
Which type of ammeter would you select to measure alternating current at this frequency?
Give reasons for your choice.

(5) With the aid of a diagram derive an equation for the balanced condition for a Wheatstone bridge.
Draw a diagram of a 'G.P.O. box' form of Wheatstone bridge and using typical or reasonable resistance values state the resistance settings which you would expect if measuring a resistance of 15 Ω.

(6) Explain what is meant by resonance in connection with a series circuit, and derive an equation giving the resonant frequency in terms of inductance and capacitance.

A series circuit has a resistance of 20 Ω, an inductance of 0·3 H and a capacitance of 80 μF.
Calculate:

(*a*) the resonant frequency,
(*b*) the power factor at resonance,
(*c*) the power dissipated at resonance,
(*d*) the voltage across the capacitor assuming the supply is 100 V r.m.s.

(7) State Kirchhoff's laws in connection with an electrical circuit. A two-core dc distribution cable *AB* 800 m long is fed at *A* at 210 V and at *B* at 200 V. If concentrated loads of 60 A and 80 A are supplied at points 150 m and 500 m respectively from end *A*, determine the current taken at *A* and at *B*, and the p.d. across each load. The resistance per core is 0·03 Ω/100 m.
Illustrate your answer with a diagram.

(8) With the aid of a circuit diagram explain how you would arrange an open circuit test on a dc separately excited generator. Sketch the shape of two open-circuit characteristics which could be obtained from such a test.
The curve of induced e.m.f. for a separately excited dc generator when run at 1200 rev/min on open-circuit is given by:

e.m.f. (volts)	12	44	73	98	113	122	127
Exciting current (amperes)	0	0·2	0·4	0·6	0·8	1·0	1·2

Determine graphically the open-circuit voltage of this machine when shunt-connected if the total field circuit resistance is 110 Ω, and the machine is running at 1200 rev/min.

(9) Sketch neatly the constructional details of a hand-operated insulation tester (commonly known as a 'Megger'), and answer briefly the following questions:

(*a*) Is the test voltage alternating or unidirectional?
(*b*) What is the usual relationship between test voltage and working voltage?
(*c*) How is the maximum voltage limited?
(*d*) Why is the maximum test current usually limited to about 5 mA?
(*e*) Does the speed at which the handle is turned affect the reading of the instrument?
(*f*) Why does the pointer not return to zero after use?

This paper included by courtesy of the East Midland Educational Union.

TEST PAPER THREE

(1) A three branch parallel circuit is fed from a 240 V, 50 Hz sinusoidal supply.

Branch (i) consists of a 60 Ω resistor,
branch (ii) consists of a 50 μF capacitor, and
branch (iii) consists of a coil whose resistance is 40 Ω and reactance 30 Ω.

Determine:

(a) the current and power factor of each branch,
(b) the supply current, the overall power factor and the total power.

(2) (a) Describe, with the aid of a circuit diagram, a direct reading dc potentiometer. Detail the process of 'standardising'.
(b) Explain briefly how the potentiometer may be used to check the calibration of a 0–1 A ammeter.

(3) (a) Explain the essential difference between lap and wave windings, showing how the choice of winding affects the current and voltage rating of a dc machine.
(b) A 6 pole, wave-wound dc generator has an armature current of 100 A at a terminal voltage of 440 V. If the flux per pole is 0·02 Wb, the rated speed 1000 rev/min and the armature resistance 0·12 Ω, calculate:

(i) the generated e.m.f.,
(ii) the total number of armature conductors.

(4) Explain, with the aid of a graph and a phasor diagram, what is meant by resonance in a series circuit. Deduce the formula for resonant frequency.
A coil of resistance 10 Ω and inductance 0·159 H is connected in series with a variable capacitor across a 240 V, 50 Hz sinusoidal supply. Calculate for the resonant condition: (a) the circuit impedance (b) the current, (c) the circuit power factor, and (d) the voltages across the capacitor and coil respectively.

(5) State the e.m.f. equation of a transformer, clearly indicating the meaning of all symbols used.
A 10 kVA, 240/1200 V, 50 Hz, single phase transformer has 60 turns on the primary winding. Ignoring the no load current, calculate:

(a) the number of turns on the secondary,
(b) the full load primary and secondary currents,
(c) the maximum flux density if the core has a cross-sectional area of 0·014 m^2.

Sketch, *on the same axes*, graphs to show how (i) iron loss (ii) copper loss and (iii) efficiency, vary as the load is varied from zero up to full load.

(6) With the aid of circuit and phasor diagrams, deduce the relationship between the phase and line voltages of a 3-phase, balanced, star-connected sinusoidal supply.

Three identical coils are connected in star to a 3-phase, 3-wire, 415 V, 50 Hz sinusoidal supply, the line current being 8 A at 0·5 power factor lagging. Calculate the total input power. If the same coils are now delta-connected to the same supply, calculate the line current.

(7) Describe briefly, with sketches, the construction of a direct reading dynamometer wattmeter, showing how deflecting, controlling, and damping torques are obtained. Draw diagrams to show two alternative methods of connection of current and voltage coil systems, and state, with reasons, which should be used with a low impedance load, and which with a high impedance load.

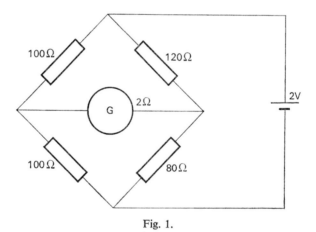

Fig. 1.

(8) The circuit in Fig. 1 represents an unbalanced Wheatstone Bridge.

(a) Using Maxwell's method, write down, *but do not solve*, the *minimum* number of equations necessary to find the galvanometer current.

(b) Calculate this current using *Thévenin's* theorem.

(9) (a) State the disadvantages, to both consumer and supplier, of low power factor industrial loads.

(b) A small inductive load takes a current of 0·75 A when connected across a 240 V, 50 Hz, sinusoidal supply. The power dissipated is 80 W. Find the value of capacitance to be connected across the supply terminals to improve the overall power factor to 0·95 lagging.

(10) State briefly the main function of an amplifying device. What are the main requirements of the following types of amplifier? Give *one* application of each.

(*a*) low frequency amplifier,
(*b*) dc amplifier,
(*c*) tuned amplifier.

This paper included by courtesy of the Northern Counties Technical Examinations Council.

TEST PAPER FOUR

(1) Two coils are connected in series. When a current of 2 A dc flows through the circuit, the p.d. across the coils are 20 V and 30 V respectively. When a current of 2 A ac at 40 Hz flows in the circuit the p.d. across the coils are 140 V and 100 V respectively.
If the two coils in series are connected to a 230 V, 50 Hz supply, calculate (i) the circuit impedance, (ii) the current, (iii) the power, (iv) the power factor.

(2) (a) With the aid of a phasor diagram, explain the circumstances in which voltages in excess of the supply voltage may occur in a series circuit containing a coil and a capacitor.
(b) What are the characteristics of resonance in a parallel ac circuit?

(3) A single-phase motor takes a current of 20 A from a 400 V, 50 Hz supply, the power factor being 0·8 lagging.
If the motor power remains unchanged, determine:

(i) the capacitance to be connected in parallel with the motor to raise the power factor to 0·9 lagging,
(ii) the additional capacitance required to raise the power factor to unity.

(4) (a) Draw a connection diagram showing how a wattmeter, current transformer, voltage transformer, ammeter and voltmeter are arranged for measuring single-phase voltage, current and power through instrument transformers in a high-pressure system.

(b) (i) Why must the current transformer secondary *not* be left open-circuited?

(ii) Why is the secondary winding of an instrument transformer usually earthed?

(5) (a) Discuss the core construction of a transformer used in a power-supply system.
(b) A 10 kVA single-phase transformer has a turns ratio of 300:23. The primary is connected to a 1500 V, 60 Hz supply. Find (i) the secondary voltage on open-circuit, (ii) the approximate values of the currents in the two windings on full-load, (iii) the maximum value of the flux.

(6) (a) Explain the constructional differences in the field coils of shunt and series wound dc machines.
(b) Sketch an end-view of a dc machine, with cover removed, having 4-poles and a cylindrical yoke.

Label the diagram showing clearly (i) the poles and their polarities, (ii) the pole shoes, (iii) the yoke, (iv) the field winding, (v) the armature, (vi) the mean path of the magnetic lines of force and their directions.

(7) (a) Explain how speed control of a dc shunt motor is obtained by variation of the field current. What are the limitations of this method?

(b) A 250 V shunt motor with an armature resistance of 0·5 Ω takes an armature current of 20 A when running on no load at 1000 rev/min. If the load torque remains unchanged find the approximate value of the armature current and the speed, if the field is weakened by 10 per cent.

(8) When a Wheatstone bridge is balanced, the four arms have the following resistances: AB = 100 Ω, BC = 40 Ω, AD = 156 Ω, CD = unknown.

(i) Calculate the value of the unknown resistance,

(ii) calculate the current through each resistance when a battery having an e.m.f. of 4 V and an internal resistance of 20 Ω is connected across AC.

(9) Assuming an expression for the power per phase show that the expression for power in terms of line voltage, line current and power factor in a balanced 3-phase system is the same for both star and delta connection.

A 3-phase 440 V delta connected induction motor takes a line current of 200 A when started on full voltage in the delta connection. Calculate the starting current and kVA when this motor is connected in star for starting.

This paper included by courtesy of the Northern Counties Technical Examinations Council.

TEST PAPER FIVE

Certain of these questions are obtained from the Union of Educational Institutions and are marked U.E.I.

(1) State Thévenin's theorem.

Two batteries, one 4 V with internal resistance 2 Ω, the other 6 V with internal resistance 3 Ω, are connected in parallel across a load resistor. The load current is found to be 0·3 A. Use Thévenin's theorem to determine the value of the load resistance.

(2) Define the terms 'time constant' and 'steady state value' as applied to the connection of a dc source to an inductive coil.

An inductive coil of resistance 10 Ω is connected to a 100 V dc supply. After 2 seconds the coil current is 3·56 A. Determine the value of the time constant of the circuit and hence the coil inductance.

(3) Explain briefly the reasons for the use of power factor correction equipment in industry.

An inductive load when connected to a 10 V ac supply takes a current of 13 A and absorbs 50 W of power. A capacitor is then connected in parallel with the load and takes a current of 8·25 A. Calculate the power factor of the inductive load before the capacitor is connected and the new power factor of the combined load after the capacitor is connected.

(4) Derive the equations connecting the line and phase currents of a star connected three-phase load.

Three coils each having a resistance of 20 Ω and reactance of 15 Ω are connected in star to a 400 V three-phase supply. Calculate (a) the line current, (b) the power supplied, (c) the power factor.

(5) Explain the meaning of 'control and damping' and how this is effected in a moving coil instrument. Explain with diagrams how a moving coil instrument may be adapted for measuring voltage and current.

Calculate the value of the additional resistors needed to set up

(a) a 15 A f.s.d. ammeter and

(b) a 100 V f.s.d. voltmeter using a moving coil instrument having a resistance of 5 Ω and a full scale deflection current of 15 mA. (U.E.I.)

(6) Describe with the aid of a diagram the principle of operation of a simple vacuum diode valve. Explain with the aid of a circuit diagram how such a diode can be used as a half wave rectifier illustrating your answer with graphs of the applied voltage and the output load current. (U.E.I.)

(7) What is meant by the term 'critical resistance' as applied to a shunt excited dc generator. Determine the approximate value of the

critical resistance of a machine having the following open circuit characteristic:

Field current	0	0·1	0·3	0·5	0·7	0·9	amperes
Generated voltage	0	27	57	86	106	118	volts

From the graph determine also the o.c. e.m.f. if the field resistance is 160 Ω.

(8) Describe and explain the principle of the Wheatstone bridge illustrating your answer with a wiring diagram. State two forms of the Wheatstone bridge used in practice. The ratio arms P, Q of a Wheatstone bridge are 1000 Ω and 10 Ω respectively and the variable resistance X connected opposite to arm Q reads 273 Ω. Calculate the resistance of the unknown resistance R connected opposite to arm P. (U.E.I.)

(9) What is the meaning of

(a) relative permittivity?

(b) electric force or potential gradient as applied to a capacitor?

Two metal plates 200 cm^2 in area are immersed in an insulating oil and are spaced 2 mm apart. The measured capacitance of the capacitor so formed is 442·5 pF. Calculate the relative permittivity of the oil.

If the capacitor is connected to a 200 V supply calculate:

(i) the charge stored on the plates,

(ii) the potential gradient.   (U.E.I.)

Answers to Problems

Answers are given to slide rule accuracy; those marked with an asterisk (*) are derived from graphical solutions and a greater tolerance should be allowed when comparison is made with answers obtained by the reader.

Chapter 1

(1) 6·67 μF, 66·7 μC   (2) 33·3 V across 8 μF, 66·6 V across 4 μF
(3) 141·4 V, 70·7 V   (4) 0·1 s, 0·1 s, 0·1 s, 0·5 s   (5) 165 ms, 10·8 V
(6) 25 ms, 25 ms   (7) 450 Ω, 10 ms   (8) 50×10^5 A/Wb
(10) 5000   (11) 1·5 μA   (12) 0·2 ms, 0·138 ms, 0·3 A   (13) 50 kΩ,
20 Ω   (15) 7·9, 0·875 pF.

Chapter 2

(1) 95·5 mH   (2) 85·6 V, 31·8 V; 0·856 A, 0·256 rad; 73·25 W, 0·968
(3) 1·06 A, 70·7 V   (4) 0·552 H   (5) 1·643 μF   (6) 7·58 A
(7) 15·4 A, 0·329 rad   (8) 1 kHz, 2, 0·314 Ω   (9) 56·4 Hz, 32 mA,
0·492 A   (10) 200 VA, 100 W, 173·2 var   (11) 619 kvar
(12) 5·17 kW   (13) 20 Ω, 14·9 A, 179 V, 358 V, 447 V
(14) 3·98 mH, 15·9 nF, 40 V   (15) 10 mV, 50 Ω, 0·1 Ω, 1 mW
(16) 0·1 A, 5 V, 5 V, 5, 1 V.

Chapter 3

(1) 2·18 V   (2) 0·232 A   (3) 3 W   (4) 0·436 A   (5) 15 Ω, 16·32 V
(6) 0·11 A, 0·67 V   (7) 83·3 mW, 81·7 mW, 98·2 per cent
(8) 23 mA   (9) 5·2 V   (10) 0·179 A.

Chapter 4

(1) 87·5 A   (2) 19·65 A, 13·5 kW   (3) line 42·2 A, phase 24·4 A; line
14·1 A, phase 14·1A   (4) 21·4 A, 0·447, 37·2 A, 6·9 kW   (5) 400 W
(6) 71 A   (7) 20·7 kW, 27·6 kVA, 18·25 kVAr, 20·9 A
(8) 7·92 A, 4·72 kW, 6·04 kVA.

Chapter 5

(1) 1·5 T, 600 turns   (2) 2 V/turn, 500 turns   (3) 2·8 A, 0·77 lagging*
(4) 25, 1·8 T   (5) 1·83 kW, 1·83 kW, 98·9 per cent   (6) 13·75 A*
(7) 99·3 per cent, 50·4 A   (8) 100 V, 10 A, 100 A   (9) 97·34 per cent, 96·9 per cent   (10) 320 V, 13·75 A*.

Chapter 6

(1) 285 V*   (2) 387·3 rev/min   (3) 242·8 V   (4) 818·1 rev/min
(5) 207·2 Nm   (6) 613 rev/min   (7) 507 rev/min   (8) 238·2 V
(9) 405 Nm   (10) 488 conductors.

Chapter 7

(5) Transformer of ratio 44·7:1.

Chapter 8

(2) Multipliers 3328·3 Ω, 6661·7 Ω, 13 328·3 Ω   (3) 24 990 Ω,
0·204 Ω   (6) 800 kW   (10) 2·77 A   (12) 145 Ω.

Chapter 9

(1) 1053 kΩ   (2) 66 646 Ω, 0·012 Ω   (3) 900 kW   (4) 2·25 Ω
(5) 42·1 Ω, 0·19 V   (6) 134 kΩ   (7) Reading 4 W, actual 3·2 W
(9) Series resistor 1300 Ω   (11) 1·5 kHz.

Chapter 10

(1) +1V   (3) 2 per cent   (5) 1000, 2·24   (6) 9·99, 0·1375, 9·8525,
10·1275   (7) 5 Ω ± 3·1 per cent   (8) 1·12 per cent, 5·07 per cent
(9) 10 mV, 16·6 µA, 0·166 µW; 1 mV, 1·66 µA, 1·66 nW
(11) 0·05 W, 500 Hz, 100·25 kHz, 99·75 kHz.

The answers given to the test papers are those obtained by the author; the appropriate examination board from whom the questions were obtained is in no way responsible for their accuracy.

Test Paper 1

(1) 174 V*   (2) 752 rev/min   (3) 0·119 A   (4) 80 mA; 503 V; 2·09
(5) 10·1 kvar   (6) 50 ms, 0·138 A   (7) 800 W; 5·55 A; 26 Ω
(8) 8650 V   (9) 2·5 Ω, 68·7 V.

Test Paper 2

(1) 10 Ω; 239 V; 23·9 A; 10·3 kW; 0·6 (2) 8·33 A; 36·9 μF; 2·9 A (6) 32·5 Hz; unity; 500 W; 306·3 V (7) A:120·4 A, B:19·6 A; 204·58 V, 198·24 V (8) 125 V*

Test Paper 3

(1) 4 A, 3·77 A, 4·8 A; unity, zero, 0·8; 10·28 A, 0·7619, 1·88 kW
(3) 452 V; 452 conductors (4) 10 Ω; 24 A; unity; 1200 V
(5) 300; 41·6 A, 8·3 A; 1·29 T (6) 2875 W, 24 A (8) 4 mA
(9) 7·478 μF.

Test Paper 4

(1) 119·6 Ω; 1·925 A; 93 W; 0·21 (3) 33·89 μF; 95·48 μF
(5) 115 V; 6·67 A, 87 A; 18·77 Wb (7) 22·2 A, 1106 rev/min
(8) 62·4 Ω; 14·84 mA (9) 115·5 A, 50·8 kVA.

Test Paper 5

(1) 14·8 Ω (2) 4·55 s; 45·5 H (3) 0·385, 0·8 (4) 9·24 A,
5130 W, 0·8 lagging (5) 5005 $\mu\Omega$, 6·662 kΩ (7) 235 Ω, 101 V
(8) 2·73 Ω (9) 5, 0·0885 μC; 100 V/mm.

APPENDIX I

The International System of Units (SI)

The three basic mechanical units of this system are those of *length*, *mass*, and *time*, being metre, kilogramme, and second respectively. All other mechanical units (excluding temperature) are derived from these and are given special names as shown:

Quantity	Symbol	Basic name	Special name
Length	l	metre	—
Mass	m	kilogramme	—
Time	t	second	—
Speed	$v = l/t$	metre/second	—
Acceleration	$a = l/t^2$	metre/(second)^2	—
Force	$F = ml/t^2$	kilogramme metre/(second)^2	newton
Energy (work)	$W = ml^2/t^2$	kilogramme (metre)^2/(second)^2	newton metre or joule
Power	$P = ml^2/t^3$	kilogramme (metre)^2/(second)^3	joule/second or watt

A system containing electrical and mechanical quantities requires a fourth basic unit, which in SI is the *ampere*. This is the current which when flowing in two wires set one metre apart sets up a force (caused by magnetic field interaction) of 2×10^{-7} newtons. All other electrical units are derived from these four units a few of the more important ones being as follows:

coulomb	ampere second
volt	joule/coulomb
farad	coulomb/volt
weber	volt second
henry	volt second/ampere
ohm	volt/ampere

APPENDIX II

Frequency Meters

Vibrating reed frequency meters contain a set of steel reeds of suitable dimensions, each one having a slightly different resonant frequency from the one next to it. A coil, which is supplied from the circuit in which the supply frequency is required, sets up a fluctuating magnetic field and the reed or reeds having a resonant frequency nearest to that being measured vibrates in sympathy with this field. If two reeds vibrate with the same amplitude the frequency lies between the resonant frequencies of the two reeds. Because of the necessity of providing a very large number of reeds to accommodate a wide frequency spectrum, this type of meter is normally only useful for a narrow band of frequencies and as such would be used as a check on a frequency having a value which fluctuates only slightly, *e.g.* the mains frequency which might require a range between 47 Hz and 53 Hz in 0·5 Hz increments, *i.e.* 12 reeds.

Electrodynamic and induction frequency meters use the principle of interaction between two magnetic fields to produce a deflection of a needle across a calibrated scale. The two fields are produced, one by a resistive circuit, the other by a reactive circuit, so that a higher frequency produces a lower current in the reactive circuit than in the resistive circuit and this produces an appropriate magnetic field and thus deflection. Conversely, if the frequency is reduced, the resistive circuit current is less than that in the reactive circuit and an opposite deflection is produced.

Frequency measuring systems contained in a single piece of apparatus and loosely referred to as frequency meters may use the heterodyne principle or gated digital techniques. In the first kind the unknown frequency is *mixed with* or *beaten against* a known standard frequency to produce a difference frequency. This is then compared with the difference frequency produced by *two* known frequencies until both difference frequencies are the same, the second being directly measurable from the known standards. In the digital frequency measuring system a number of current or voltage fluctuations at the unknown frequency are *sampled* for a predetermined period and the fluctuations used to produce pulses which are then counted by standard electronic counting techniques.

Index